D0927496

The Human Microbiome

The Human Microbiome

ETHICAL, LEGAL AND SOCIAL CONCERNS

Edited by Rosamond Rhodes, Nada Gligorov,
and Abraham Paul Schwab

OXFORD
UNIVERSITY PRESS

Oxford University Press is a department of the University of Oxford.
It furthers the University's objective of excellence in research, scholarship,
and education by publishing worldwide.

Oxford New York
Auckland Cape Town Dar es Salaam Hong Kong Karachi
Kuala Lumpur Madrid Melbourne Mexico City Nairobi
New Delhi Shanghai Taipei Toronto

With offices in
Argentina Austria Brazil Chile Czech Republic France Greece
Guatemala Hungary Italy Japan Poland Portugal Singapore
South Korea Switzerland Thailand Turkey Ukraine Vietnam

Published in the United States of America by
Oxford University Press
198 Madison Avenue, New York, NY 10016

Library of Congress Cataloging-in-Publication Data

The human microbiome : ethical, legal and social concerns / edited by Rosamond Rhodes, Nada Gligorov, and Abraham Paul Schwab.
pages cm
Includes index.
ISBN 978-0-19-982941-5 (hardback : alk. paper) — ISBN 978-0-19-932215-2 (updf) 1. Medical microbiology—Research—United States. 2. Human body—Research—Moral and ethical aspects—United States. 3. Bacterial genomes—Research—United States. 4. Human body—Law and legislation—United States. 5. Medicine—Research—United States. I. Rhodes, Rosamond. II. Gligorov, Nada. III. Schwab, Abraham P.
QR46.H85 2013
616.9′041—dc23
2013000230

9 8 7 6 5 4 3 2 1
Printed in the United States of America
on acid-free paper

To our spouses
Joe Fitschen, Stephen Krieger, Liz Schwab
for their love, encouragement, and contribution to our healthy microbiomes

CONTENTS

ACKNOWLEDGMENTS

For two years, the Microbiome Working Group members participated in the discussions that resulted in this book. Participants presented material from their specialties to inform the group, shared comments, raised questions, and offered suggestions on policy recommendations and on the several versions of the various chapters.

MICROBIOME WORKING GROUP PARTICIPANTS

Jody Azzouni, PhD, Tufts University
Mary Ann Baily, PhD, Hastings Center Fellow
Stefan Bernard Baumrin, PhD, JD, The Graduate Center, CUNY
Keith Benkov, MD, Icahn School of Medicine at Mount Sinai
Martin J. Blaser, MD, New York University Medical School
Barbara Brenner, PhD, Icahn School of Medicine at Mount Sinai
Joseph W. Dauben, PhD, The Graduate Center, CUNY
William J. Earle, PhD, The Graduate Center, CUNY
Lily E. Frank, MPhil, The Graduate Center, CUNY
Nada Gligorov, PhD, Icahn School of Medicine at Mount Sinai
Joseph Goldfarb, PhD, Icahn School of Medicine at Mount Sinai
Kurt Hirschhorn, MD, Icahn School of Medicine at Mount Sinai
Rochelle Hirschhorn, MD, New York University Medical School
Ian R. Holzman, MD, Icahn School of Medicine at Mount Sinai
Debbie Indyk, PhD, Icahn School of Medicine at Mount Sinai
Ethylin Wang Jabs, MD, Icahn School of Medicine at Mount Sinai
Douglas P. Lackey, PhD, The Graduate Center, CUNY
Daniel A. Moros, MD, Icahn School of Medicine at Mount Sinai
Sean Philpott, PhD, Union Graduate College
Matthew E. Rhodes, PhD, Pennsylvania State University
Rosamond Rhodes, PhD, Icahn School of Medicine at Mount Sinai
Lynne D. Richardson, MD, Icahn School of Medicine at Mount Sinai
Henry S. Sacks, MD, Icahn School of Medicine at Mount Sinai
Abraham P. Schwab, PhD, Indiana University, Purdue
Rhoda Sperling, MD, Icahn School of Medicine at Mount Sinai
Brett Trusko, PhD, Icahn School of Medicine at Mount Sinai
Arnulf Zweig, PhD, University of Oregon (emeritus)

Our work was funded by 1R01HG004856-01 as a component of the Human Microbiome Project (HMP), a National Institutes of Health (NIH) Common Fund initiative.

We thank Dianne E. Hoffmann, JD, MS, and the University of Maryland Law School for inviting Rosamond Rhodes and Lily E. Frank to participate in their NIH HMP meetings on Federal Regulation of Probiotics. We are grateful for all that we have learned from those meetings.

In addition, we want to acknowledge the constant support of Karen Smalls, who took care of the numerous details involved in making this project work. We also want to thank Rebecca Ellis and Marty Rotenberg who worked as summer interns and assisted in the production of the glossary and review of the final drafts of the manuscript.

Introduction

LOOKING BACK AND LOOKING FORWARD

Rosamond Rhodes

Scientific knowledge about the microbiome is already filtering into the public's awareness. On television, Jamie Lee Curtis charms viewers into making Activia a regular part of their diet (pun intended). On Web MD, Kathleen Doheny announces that "Fecal Transplant May Treat Stubborn C. diff: Study Shows Procedure Can End Symptoms of Diarrhea" (2011), a topic that TV comic Stephen Colbert also mentioned with disgust. In the popular television series *NCIS*, crime scene investigators use the bacteria in sewage to solve a mystery, while investigators on *CSI Miami*, another television series, use the microbiome from a computer keyboard to solve one of theirs. And the *New York Times* publishes a couple of articles with intriguing titles by Carl Zimmer, "How Microbes Defend and Define Us" (2010) and "Our Microbiomes, Ourselves" (2011). The science is intriguing, and the applications could be revolutionary.

According to Nobel Laureate, geneticist, and microbiologist Joshua Lederberg, the cohabitation of the human genome and the genomes of the bacteria and viruses that occupy our skin, mucous membranes, and intestinal tract makes up the microbiome. Lederberg claims that understanding the microbiome is central to understanding the dynamics of health and disease (Lederberg 2003). Research on the human microbiome is in its infancy. Yet, this new area of biological science promises to launch a new age of pharmacogenomics, personalized medicine, and personalized nutrition—diets and treatments that will be chosen for individual patients based on features of their microbiome (Bylund et al. 2007; Shay 2007; Vasquez 2004). Improved understanding of the human microbiome also promises to advance systems and statistical modeling that can be used to track the spread and transformation of microorganisms and to predict health outcomes. Such information can be translated into public health measures that will benefit the population at large (Hodge and Brown 2005; Holford et al. 2005).

The concept of a human microbiome also has dramatic implications for how we think of ourselves because it challenges the view of ourselves as atomistic individual organisms. This new concept may recast us as an amalgam of us and them. Or it may encourage us to think of ourselves as superorganisms that include our own cells and a fluctuating set of bacteria and viruses that share our body space and sometimes even enter our genome. As Jeffrey Gordon and colleagues explain, the collective genomes of the microbiome endow us with physiologic capacities that we have not had to evolve on our own, and so may actually have been instruments in accelerating our evolution. In this light, human function is not simply determined by our own physiology, but includes the microbiota that help regulate energy balance, digest food, and modulate the immune system (Gordon et al. 2005). Further, because the viruses and bacteria of our microbiome grow and evolve in cycles of 20 minutes or less and exchange genes laterally—that is, within and between various species—the combined human–microbiome self is more dynamic and more interactive than we are used to thinking of ourselves as being (Lederberg 2003). In the sense that our physical bodies and the space occupied by our microbiome are hard to delineate, and that the human microbiome genome cannot be entirely separated from the human genome, we are coexistent rather than independent beings.

This new understanding of our coexistence with the microbiome tells us that human evolution is not just human history, but the story of our interaction with the viruses, fungi, and bacteria that inhabit us. Learning more about the microbiome is likely to change the way medicine is practiced. It may also have implications for our social and legal systems. There may also be important implications for how we conceive and address the ethics of medicine and biomedical research. The goals of our project were, therefore, to identify the ethical, legal, and social implications raised by human microbiome research, in order to provide insight and guidance for scientists engaged in the work and members of our society who will be asked to cooperate in studies and live with the consequences. At the least, we would like to initiate conversations about these issues.

In producing this volume, our working hypothesis was that learning about the human microbiome will expand the scope of ethical reflection even when it does not raise entirely unique issues. We tried to do two things at once: (1) to locate our understanding of the microbiome within the existing rich and intricately textured social fabric of concepts, history, regulations, and practices, and (2) to reexamine common beliefs, policies, and positions in multiple domains from the new perspective afforded by our developing scientific appreciation of the human microbiome. Some issues raised by study of the human microbiome are very similar to issues encountered in other domains. Other issues may turn out to be significantly different from those that arise elsewhere. In those cases, we have tried to consider just how novel these factors are and how they should be addressed. Accordingly, we have identified relevant models and points of comparison for grounding our response to the ethical, legal, and social issues that emerge from work on the human microbiome. Seeing issues from this new vantage point moved us to ask and answer questions about the conceptual foundation of accepted principles and distinctions and about the relative importance of previously accepted commitments. This activity provided an opportunity for rethinking and reevaluating some positions previously accepted in similar cases. It encouraged us to try to expose views that have been influenced by ideology, to lay bare oversights of previous analyses, and to illuminate issues that remain obscure and troubling.

We are aware that we are exploring a domain full of uncertainty. Scientific knowledge related to the microbiome is being developed at an amazing pace, but we are still in the very early stages of understanding microorganisms, many of which we cannot yet culture independently. We cannot know what we will learn. We cannot know whether what we now think will turn out to be entirely accurate. At many points we are making predictions, but we are uncertain about what will actually happen and whether the effects will turn out to be beneficial or harmful. At the very least, everything we say has to be viewed as tentative and limited by the context of today's knowledge.

The Project Collaborators

With the support of a National Institutes of Health (NIH) Human Microbiome Project grant, we assembled an interdisciplinary team of 27 health professionals, scientists, and scholars from the humanities and social sciences to explore issues raised by human microbiome research. Through an intense process of mutual education, group discussion, consensus formation, writing, and critiquing, we composed the chapters that make up this volume. The collaborators on this project were invited to participate based on the relevance of their expertise and their availability to work together frequently and in close proximity. Each of the participants recognized the need for the conceptual work that this project aimed to produce, each had a vested professional interest in clarifying the key issues, and each actively contributed to the positions presented in this volume.

Several members on our team have significant experience in medical and research ethics. Our collaborators include the current and past chairs of the Mount Sinai Hospital's Ethics Committee, three additional ethics committee members, past and present members and chairs of national society ethics committees, and several faculty members who have served on institutional review boards (IRBs) and institutional animal care and use committees (IACUCs). Several team members have also presented papers and published on bioethical issues.

Many members of the faculty at the Icahn School of Medicine at Mount Sinai (ISMMS) faculty are engaged in clinical research, innovative treatment, and public health initiatives that are directly relevant to this project. Those working on this project come from an array of departments that are critical to our collaboration: Genetics and Genomic Sciences, Obstetrics and Gynecology, Pediatrics, Medicine, Community Medicine, and Emergency Medicine. They bring needed expertise in biobanking, population genetics, inflammatory bowel disease, sexually transmitted infections, infectious disease, and public health. Some of our ISMMS collaborators have also been deeply engaged in studies to advance our understanding of the need for community consultation, how the process should proceed, and how to implement ethically responsible community engagement.

In addition, we have invited researchers from outside the ISMMS to participate in this project specifically to augment our team's scientific and clinical expertise with their knowledge of microbiology and the human microbiome. We also included academics with expertise in the humanities and social sciences. Thus, we collected a group of scholars from the region with expertise in moral and political philosophy, philosophy of law, economics, and the history and philosophy of science. Some participants were chosen for their previous work in applied ethics and relevant areas of bioethics. Others were invited precisely because they have not worked in bioethics and so are naive with respect to the biases and dogmas that may affect those who work in the field.

Our Method

We aimed at producing a useful introduction to the new science of the human microbiome and to identify the kinds of social issues that scientific advances would raise. Our aim was to present thoughtful opinions informed by a solid understanding of the science, the clinical implications, and the ethical, social, and legal concepts involved. To be a valuable product for our society, we wanted to avoid the common public opinion survey approach of collecting and reporting on uninformed reactions. Instead, our method was to draw on the expertise that our working group could contribute and build on that expertise to develop well-informed positions. Our method, therefore, involved starting with a process of group education in preparation for the meetings, which were seen as an opportunity to reason together.

To that end, each of our particular topics was addressed theoretically, with historical background and the most current available information on microbiome research and advances. Every issue that we scrutinized, such as personal identity, property, privacy, and the ethical conduct of research, was challenging. The underlying concepts and their philosophical and practical implications had to be well understood before we could compare activities in other domains with those related to human microbiome research. Some of the topics that we addressed are general theoretical issues with broad implications.

Some, such as informed consent and vulnerability, bear directly on human subject research. Others, such as privacy and property, needed to be addressed because of the ethical, legal, and social issues raised by the scientific advances.

Furthermore, we wanted to provide readers with the conclusions of our deliberations. So, wherever it seemed appropriate, we developed public policy suggestions. The policy recommendations that we include at the end of chapters 2 through 7 are consensus statements that reflect the perspectives of our broadly interdisciplinary Microbiome Working Group. We offer them with the caveat that these recommendations were promulgated in the face of a good deal of uncertainty about the science and ignorance of what might develop.

When we began work on this project, the entire project Microbiome Working Group convened for preliminary discussions and to develop a list of chapter topics. Thereafter, we convened for monthly full-day meetings for nearly two years. Each working group member was assigned to at least one chapter team. The teams convened for their own meetings in which they defined the scope of their investigation, developed plans for distributing their work, and drafted their preliminary approach to their topic. Chapter leaders Rosamond Rhodes, Nada Gligorov, and Abraham Schwab were each given the responsibility for the successful completion of concept-based chapters, and Lily Frank, our project coordinator, was given the responsibility of chapter leader for the scientific and historical background chapter. Chapter leaders collaborated with faculty experts on their team. As project coordinator, Lily Frank also helped to identify a collection of historical literature and the most up-to-date scientific and theoretical literature and she posted it on a project intranet website for group education. Drawing upon their research and the chapter team's preliminary conversations, the members of each team presented the multiple specific issues related to their chapter to the full project faculty and led day-long working group discussions.

Based on these initial working group conversations, each chapter team produced a working chapter draft that was reviewed and discussed at subsequent project working group meetings. The schedule of meetings allowed sufficient time for careful preparatory reading of the background materials and chapter drafts and assured discussion of all of the aspects of the topics that required deliberation and debate. Reflecting our overarching goals, each discussion aimed at answering two questions: (1) What should our ethical, social, and legal stance be with respect to the issue? (2) Are the implications for the human microbiome significantly different from our thinking with respect to relevant comparison cases? How so, and why?

In the working group discussions we strove to achieve a consensus on policy recommendations by trying to frame positions that reasonable people could endorse. Achieving agreement on these issues required an open exchange of ideas with participants freely questioning one another's views. In some cases this was a complicated process that involved clarifying positions, refining points, and sorting out relevant distinctions such as the degree of risk involved in a research study or the feasibility of implementing some measure. To the extent that disagreement remained, participants tried to reframe the issues and clarify the outstanding points of contention. In sum, the process led us to achieve a significant convergence of opinions.

The Scope of this Volume

In this volume we focus on the most general features of human microbiome research and try to identify broad ethical, legal, and social issues related to the microbiome. We begin with (1) a background chapter that provides critical information about microbes and their place in our world and human history, and information about today's technology in studying them. Most significantly, chapter 1 explains that microbes are not the enemy that has to be eliminated but instead that microbes are essential to good health and human life. The following chapters then address several distinct ethical, legal, and social domains in which work on the human microbiome is likely to have significant implications: (2) personal identity and self-identity, (3) property and commercialization, (4) privacy and confidentiality, (5) human subject research, (6) sample banking and biobanking, and (7) population and public health research. Each of these topics involves conceptually interesting issues and insights, historically relevant examples, and information about germane regulations and legal cases. Without trying to summarize the content of the chapters, we can explain the salience of these particular issues and why they are intriguing.

Personal Identity: "I" used to be a rather simple term, and everyone knew what "I" meant. Now "I" refers to a moral agent, and the subject of my consciousness, and my phenotype, and my genotype, and my microbiome comprised of critters that are not me, some of which come and go. In different contexts different concepts of identity are relevant. Ethically, only some humans are moral persons, because only they can be held responsible for their actions (e.g., young children cannot be). As a victim of disease, my body is my identity. As a vector of disease, my microbiome is part of my identity. Although personhood and identity have never been simple concepts, as we learn more about ourselves as a network of us and them, we will have to rethink our concepts of personal identity and normalcy and what they imply (Gordon et al. 2005).

In the same way, it previously seemed clear and accurate to speak of good and bad bacteria and to indicate good and bad by their effects on human health. As research on the microbiome continues, questions of good, bad, and even normalcy will pose special challenges. Changes in the human microbiome may make it difficult to fix ranges of normalcy or even to determine what is normal for any individual. Further, the microbiome impacts the health of the human organism, but the effects of the microbiome are determined by the combined characteristics of the specific combination and proportions of microbes together with the specific genetic characteristics of the human host. Thus, a particular microbe might have positive effects in one human and negative effects in another related to either individual genetic factors or the other microbes comprising the microbiome. This will complicate efforts to identify the "normal."

Our understanding of the human microbiome and its interaction with the human body also has implications for how we conceptualize both personhood and personal identity. Personhood is usually defined in terms of essential and distinguishing characteristics. In this sense, if race distinguishes our susceptibly and resistance to disease and responsiveness to treatments, race is a part of our identity (Bamshad and Guthery 2007). Furthermore, the microbiome that lives in and on us can also be unique. In that sense, can the microbiome be incorporated into how we define ourselves as persons, and hence, considered constitutive of our personhood?

Property: In the philosophical literature, we see a radical diversity of views on property and the ethical, legal, and social questions about who owns what. Some see private property ownership as a fundamental natural right. Others see private property as an entirely social creation that stipulates what kinds of things may be privately owned and what is to be considered as property in common. On this view property rights arise out of a civil society's legal systems and institutions for the adjudication and enforcement of property claims. Clearly, these are radically different views. Acceptance of one or the other of these positions on property can have a huge impact on the advance of microbiome research.

Property issues have recently become controversial in cases involving genetic samples (Skloot 2006; Sobel and Wolman 1999). Because of these ongoing disputes, questions about property rights in biobank and sample bank deposits will have to be answered. If the human microbiome is a part of us, does that give us powerful property rights to it? Is the human microbiome the product of our labor in some sense because our body is its host? Again, the answers to these questions may incline people to see property rights in their microbiome. According to some views on property, however, under some circumstances the government can mandate invasions into (e.g., vaccinations) and uses of (e.g., compelled military service) our bodies. Can the government rightfully demand that we contribute samples of our genome or microbiome to research or public health activities? When something that is learned from the study of the microbiome turns out to have commercial value, who should reap the profits? If someone's microbiome is unique and valuable, whose property is it? And are microbes extracted from the microbiome property in the sense that they are subject to Food and Drug Administration (FDA) regulations, patent laws, and trade regulations? Although we have not yet seen lawsuits involving the human microbiome, there have recently been a few important legal rulings related to biological samples and genetic and genomic research that will be relevant to the human microbiome and that suggest policy positions.

Privacy: The borders of privacy conform with the boundaries of bodies and objects. The borders of privacy coincide with personal enclosed spaces such as my body, my home, behind the closed door of my bedroom, inside my own diary, and inside my own thoughts. With few exceptions, we generally believe that no one may enter my private domain without my permission, and governmental intrusions require very serious justification.

Confidentiality, however, is different from privacy (Francis et al. 2008). Confidentiality is an important professional responsibility for clinicians and some other professionals (e.g., lawyers, accountants, priests). In these professions an artificial space is created for safeguarding and sharing information. Within the recognized boundaries, information disclosed by those seeking professional services may be shared to promote the client's interests. Outside of those boundaries, disclosed information may not be divulged unless the client explicitly permits the information to be conveyed to designated others. For example, in medicine, patients expect that their medical history, diagnosis, and prognosis will only be shared among the legions of health professionals who need that information for providing their care 24 hours a day, 7 days a week. Beyond those minions who need access, patients reasonably expect that the secrets of their bodies will not be divulged unless they explicitly direct it to be shared, for example, with family members, insurers, or teams of lawyers. The assurance of confidentiality is critical for the practice of medicine because without that expectation patients would be reluctant to share information and, hence, less likely to get the medical care that they need.

To the extent that medicine and other professions are social creations, the limits of confidentiality are social artifacts. The borders of privacy follow the contours of bodies and objects. In contrast, the boundaries of confidentiality are artificially constructed based on factors related to specific professions. In medicine, confidential information is shared based on the need to know. Thus, when information is disclosed to one's doctor, the doctor may share that information with other health professionals who will be involved in that patient's care on a need-to-know basis. Physicians involved in the care of a patient may view medical records or discuss the diagnosis and treatments, as well as relevant information about the patient's life and behavior, all without a violation of confidentiality.

Although the concepts of privacy and confidentiality are frequently conflated and the distinctions elided, considering the human microbiome may reveal that the differences between them are critical. When we reexamine issues of privacy and confidentiality from the perspective of research on the microbiome, we will have to confront a number of questions. Which information should be protected, and why? Should the materials in biobanks and sample banks be considered under our strict conception of privacy or the somewhat broader limits of confidentiality? Many research ethics discussions come down strongly on the side of treating research samples with privacy protections (Annas 2003; Barnes and Krauss 2001; Clayton 1995; Kelsey 1981; Prainsack and Gurwitz 2007; Roche and Annas 2006; Woodward 1997). Is this thinking and the regulations that follow from that assumption conceptually coherent? Would the safeguards of confidentiality be a more reasonable standard for medicine and biomedical research? How would a confidentiality-based standard be different from a privacy-based standard? Is the contribution to the advance of science a sufficient justification for sharing information? Which policies for sharing samples would reflect reasonable concerns? Recently, particularly since the 1996 Health Insurance Portability and Access Act (HIPAA), a great deal of attention has been paid to the protection of privacy. This has had dramatic and controversial effects on medical care and human subject research. Do those protections serve the common good? Do HIPAA protections reflect the positions that reasonable and informed people would endorse?

Learning about the human microbiome will raise further important privacy questions with implications in matters of personal liberty. If 95% of my feces are microbiome, and my microbiome may tell the story of where I have been and with whom I have associated, should law enforcement agents have access to my microbiome in the same way that they are free to collect DNA samples that I inadvertently leave behind, for example, on a cup from which I drank?

Human Subject Research: Research ethics first received significant international attention during the Nuremberg Trials, and the 1947 Nuremberg Code began the formal march toward the 1964 World Medical Association Declaration of Helsinki and the development of our current research standards and regulations (Faden and Beauchamp 1986; Jonsen 1998; the Nuremberg Code 1947; World Medical Association Declaration of Helsinki 1964). The second half of the twentieth century was also the period of the assertion of civil rights, women's rights, and the rights of people with disabilities, all movements argued in terms of autonomy. It is no surprise, therefore, that today's research regulations are informed by the concept of autonomy, typically expressed in terms of "respect for persons." These include the Belmont Report (National Commission 1979) and the Common Rule that was developed from it, more formally known as the U.S. Code of Federal Regulations (Federal Policy 1991), as well as the International Ethical Guidelines for Biomedical Research Involving Human Subjects (CIOMS 1993). This commitment to autonomy gives primacy to informed consent and the emphasis on privacy in contemporary research ethics. Autonomy and the autonomy-related issues of informed consent and privacy will have to be addressed in the study of the human microbiome. At the same time, in 2011 the Department of Health and Human Services (HHS) posted an advance notice of proposed rulemaking (ANPR),

"Human Subjects Research Protections: Enhancing Protections for Research Subjects and Reducing Burden, Delay, and Ambiguity for Investigators," for comments. This posting indicates that the HHS recognizes that our rules for conducting research should be improved. The perspective of human microbiome research could contribute to that effort by helping to reshape the current rules and adjusting them to reflect a better balance of important human values.

Our appreciation of the importance of autonomy originates in the Enlightenment and the social contract theories advanced by Thomas Hobbes, John Locke, and Jean Jacques Rousseau, which have informed our understanding of the modern liberal democratic state. In each of these theories, the authority of the state is seen to rest on the will of its free and independent citizens. Similarly, in research ethics, the authority of researchers to employ human subjects in their studies derives from the subject's informed consent. The decision to participate in a study is taken as endorsement of the research goals and as free acceptance of the burdens and risks involved as being reasonable under the circumstances.

In keeping with the twentieth century's focus on autonomy, a good deal of the NIH Human Genome Project work on the ethical, legal, and social implications (ELSI) of research on the human genome has focused on individual liberty and autonomy-related concerns (McGuire et al. 2008). For example, *From Chance to Choice: Genetics and Justice*, the prominent book by Allen Buchanan, Dan Brock, Norman Daniels, and Daniel Wikler, supported by an NIH ELSI grant, addressed the personal choices that individuals might make about using genetic information to advance their personal goals. The authors argued in terms of autonomy for supporting individuals' decisions to seek access to treatments that would bring them up to a benchmark of "normal species function." And again, they argued in terms of autonomy against supporting individuals' choices for employing genetic information in ways that would deprive others of "the right to an open future" (Buchanan et al. 2000).

Although this focus on autonomy and the related concepts of liberty rights and their limits is certainly relevant, the literature on the ethical, legal, and social implications of the human genome study has largely overlooked the other face of social contract theory, the obligations of the subject (Gray 2000). At this stage in the development of the science, however, researchers have expanded their focus from the diagnosis of individual genetic anomalies to the broader issues of learning to understand the human genome and the interactions of the microbiome. These studies will require broad participation from the public to provide the material for biobanks and sample banks (Lee 2005). These new areas of research promise advances in knowledge that are expected to ultimately advance both personalized medicine and public health.

These efforts will raise the issue of whether or not it would be fair to expect everyone to contribute to the socially valuable research enterprise. In that some genetic and microbiome research is supported by public funds, we all do contribute our financing to the efforts. Should we also contribute bits of ourselves, our genomes, and our microbiomes? If we do not give of ourselves while expecting to benefit from what is learned, will we be free-riders on the altruism of others? Will we be treating others unjustly? To answer these questions we will need to clarify our understanding of the responsibilities of the members of our society.

Briefly, social contract theory accepts the inevitable diversity of individuals' values and allows that citizens should be free to pursue what they see as a good life. Theoretically, because societies further the core of values most people share (e.g., the desire for life, security, comfort, liberty), citizens freely agree to contribute in furthering the social good so long as others do so as well. In this way, the government's authority derives from the commitment of each individual. This reciprocal commitment of citizens allows society to provide the social benefits that we all want and that individuals alone could not provide for themselves: a reliable legal system, police and fire protection, education, safe airways, roads, clean water, sewage systems, medical facilities, and so forth. Yet, securing such benefits requires each individual to relinquish some liberties (e.g., taxes, freedoms) to achieve these broadly valued goals. Thus, the social fabric of our shared commitments is made up of the warp and weft of these two values: individual liberty and social good.

Biobanks: When it comes to biomedical science, it is easy to appreciate that everyone wants to avoid diseases and their burdens and, when they cannot be avoided, everyone would want efficient and effective means to overcome the diseases and their burdens. Having such treatments available will require significant scientific advances, which, in turn, will require a good deal more basic science and translational research. In turn, advancing scientific understanding of the human genome and microbiome will involve human subject research that utilizes biobanks and sample banks. In due course, biobank and sample bank studies are expected to advance health care through translational research and to produce personalized medicine (Lunshof 2005; McCarthy et al. 2005; Shabo 2005). These scientific efforts will require the participation of yet more individuals. The emerging needs for both more research and more general cooperation in research raises questions about the rules and governance for operating biobanks and sample banks. Who should be given access to the biological samples? Which information should be shared? Which projects should the samples support? Who should make these decisions?

Population and Public Health Research: Similar issues will arise in population and public health research (Hodge and Brown 2005; Holford et al. 2005). These issues will be particularly important now, given that the world has become dramatically more crowded with the human population now over 7 billion, up from 2.6 billion in 1950, and populations more and more concentrated in urban centers (GeoHive 2012).

The microbiome is a factor in many diseases. Efforts to protect and promote public health require disease surveillance and tracking. Data collection on disease outbreaks, births, and deaths is and will be critical in the design and implementation of effective means of disease prevention and outbreak control. Following disasters, the threat of microbe-related diseases can be particularly significant. Effective public health measures, and more local quality assurance and quality improvement efforts, should be based on data. But gathering the needed personal information may infringe upon privacy and, particularly in the context of an emergency, make prior IRB review or informed consent unfeasible (Perlman 2008).

Most research exposes only individual subjects to risks and harms. In studies of drugs and devices, individual subjects who are exposed to risks and harms are typically also possible beneficiaries of the intervention. As we begin to study the effectiveness of personalized medicine, again, we will be focused on providing boons to individuals.

Yet, as research on the human microbiome leads to studies of probiotic and phage therapies, the risks to the public as well as research subjects will have to be considered. When we contemplate altering the human microbiome by inducing changes in naturally occurring collections of microbes and viruses, we should be aware of the fact that people are both potential vectors and victims of disease (Francis et al. 2005; Smith et al. 2004). Our individual genetic makeup, together with the collections of microbes and viruses that live on us and in us, makes us more or less susceptible or resistant to disease. Furthermore, microbial agents that are beneficial to some individuals may be toxic to others. And because probiotics that help some may be communicated to others and harm them, we will have to consider how to evaluate and assess the public health risks associated with studied interventions. In this domain it is useful to consult public health ethics models of how potential risks are evaluated and how restrictions on liberty are justified for mandatory vaccination and quarantine (Nuffield Council on Bioethics 2007).

Reading this Book

Although we are just starting to understand the human microbiome, and very little of that understanding is employed at this point, it is important to think about issues in advance so that we can try to avoid at least those pitfalls that can be identified now. Examining today's microbiome research in the context of the flow of scientific revolutions provides us with an opportunity to learn from the past. Looking back at history while looking into the future encourages us to compare the consequences of previous paradigm shifts and contrast successful approaches to problematic responses. One lesson from history is that it can be useful to try to discern ethical issues and social concerns at the cusp of a transformation rather than waiting for disasters to occur and then trying to cobble together hasty responses.

This book presents our working group's collective contemplation on issues related to human microbiome research. The volume constitutes our attempt to initiate an ongoing discussion of the ethical, legal, and social issues related to human microbiome research. It also lays out some of the concepts that will be useful in considering how to respond to the novel questions created by the massive data, information, and communication issues that these scientific and technological advances will bring in their wake.

This is a multidisciplinary collaborative work by 27 contributors, each with a different academic background and a somewhat different perspective. Although the ideas in each chapter reflect the ideas and discussion contributions from all of us, each chapter has its own group of authors who contributed to writing and revising it. The chapter team leader is listed as the first author; the other chapter team authors are listed in alphabetical order.

This volume is intended to be both a coherent book and a resource for those interested in particular issues related to human microbiome research. As such, it can be read from cover to cover or read by chapter or section. To accommodate the selective reader, the table of contents lists the topics covered. After the background chapter, each chapter concludes with a list of policy recommendations, consensus positions endorsed by the full Microbiome Working Group. At the end of each chapter is a list of references for that chapter. An index to the entire volume and a glossary are intended to further facilitate the volume's use.

Because we want each chapter to be a self-contained independent entity, small amounts of material are occasionally repeated where the information is needed. This degree of planned redundancy may also serve as useful reminders in a book that spans a broad intellectual landscape.

References

Annas, G. 2003. HIPAA regulations: a new era of medical-record privacy? *New England Journal of Medicine* 348(15): 1486–1490.

Bamshad, M. and S.L. Guthery. 2007. Race, genetics and medicine: does the color of a leopard's spots matter? *Current Opinions in Pediatrics* 19: 613–619.

Barnes, M., and S. Krauss. 2001. The effects of HIPAA on human subject research. *BNA's Health Law Reporter* 10(26): 1026–1034.

Buchanan, A., D.W. Brock, N. Daniels, and D. Wikler. 2000. *From Chance to Choice: Genetics and Justice.* Cambridge, UK: Cambridge University Press.

Bylund, D. B., M.E. Kozisel, M.D. Montgomery, and A.L. Reed. 2007. Today's research, tomorrow's health: focus on pharmacogenomics. *Personalized Medicine* 4(3): 363–367.

Councils for International Organizations of Medical Sciences. 1993. *International Ethical Guidelines for Biomedical Research Involving Human Subjects (CIOMS Guidelines).* http://www.codex.uu.se/texts/international.html Accessed April 3, 2013.

Clayton, E.W. 1995. Panel comment: why the use of anonymous samples for research matters. *Journal of Law, Medicine and Ethics* 23: 375–377.

Code of Federal Regulations. 1991. Federal Policy for the Protection of Human Subjects The Federal Policy for the Protection of Human Subjects, (the "Common Rule") Title 45, Part 46. *Protection of Human Subjects*. Rockville, MD: U.S. Department of Health and Human Services. http://www.wma.net/en/30publications/10policies/b3/17c.pdfhttp://www.hhs.gov/ohrp/humansubjects/guidance/45cfr46.html Accessed April 3, 2013.

Faden, R.R., and T.L. Beauchamp. 1986. *A History and Theory of Informed Consent*. New York: Oxford University Press.

Francis, L.P., M.P. Battin, J.A. Jacobson, C.B. Smith, and J. Botkin. 2005. How infectious diseases got left out--and what this omission might have meant for bioethics. *Bioethics* 19(4): 307–322.

Francis, L.P., J.A. Jacobson, C.B. Smith, and M. P. Battin. 2008. Privacy, confidentiality, or both? *ASBH Exchange* 11(2): 1, 8–9.

GeoHive. 2012. Global data. http://www.geohive.com/ Accessed August 23, 2012.

Gordon, J.I., R.E. Ley, R. Wilson, E. Mardis, J. Xu, C. Fraser, and D.A. Relman. 2005. Extending our view of self: the human gut microbiome initiative. http://www.genome.gov/Pages/Research/Sequencing/SeqProposals/HGMISeq.pdf. Accessed April 2, 2013.

Gray, J. 2000. *Two Faces of Liberalism*. Cambridge, UK: Polity Press.

Hodge Jr., J.G., and E.C. Fuse Brown. 2005. Exchanging genetic data for public health practice and human subjects research: implications for health practitioners. *Personalized Medicine* 2(3): 259–268.

Holford, T.R., A. Windermuth, and G. Ruano. 2005. Personalizing public health. *Personalized Medicine* 2(3): 239–249.

Jonsen, A. R. 1998. The ethics of research with human subjects: a short history. In Albert R. Jonsen, Robert M. Veatch, and LeRoy Walters (eds.), *Source Book in Bioethics: A Documentary History* (pp. 5–10). Washington, DC: Georgetown University Press.

Kelsey, J. L. 1981. Privacy and confidentiality in epidemiological research involving patients. *IRB* 3(2): 1–4.

Lederberg, J.. 2003. Of men and microbes. *Frontiers of the New Century*, Summer: 53–55.

Lee, S.S. 2005. Personalized medicine and pharmacogenomics: ethics and social challenges. *Personalized Medicine* 2(1): 29–35.

Lunshof, J. 2005. Personalized medicine: how much can we afford? A bioethics perspective. *Personalized Medicine* 2(1): 43–47.

McCarthy, C.A., R.A. Wilke, P.F. Giampietro, S.D. Webrook, and M.D. Caldwell. 2005. Marshfield Clinic Personalized Medicine Research Project: design, methods and recruitment for a large population-based biobank. *Personalized Medicine* 2(1): 49–79.

McGuire, A.L., T. Caulfield, and M.K. Cho. 2008. Research ethics and the challenge of whole-genome sequencing. *Nature Reviews: Genetics* 9: 152–156.

National Commission for the Protection of Human Subjects of Biomedical and Behavioral Research. 1979. *The Belmont Report: Ethical Principles and Guidelines for the Protection of Human Subjects of Research*. Washington, DC: U.S. Government Printing Office. http://hhs.gov/ohrp/humansubjects/guidance/belmont.html. Accessed July 21, 2011.

Nuffield Council on Bioethics. 2007. *Public Health: Ethical Issues*. London: Nuffield Council on Bioethics.

Nuremberg Code. II Trials of War Criminals Before the Nuremberg Military Tribunals Under Control Law No. 10, 181-83. U.S. Government Printing Office. 1946–1949. http://www.hhs.gov/ohrp/archive/nurcode.html Accessed, April 3, 2013.

Perlman, D. 2008. Public health practice vs research: implications for preparedness and disaster research review by State Health Department IRBs. *Disaster Medicine and Public Health Preparedness* 2(3): 185–191.

Prainsack, B. and D. Gurwitz. 2007. "Private fears in public places?" Ethical and regulatory concerns regarding human genomic databases. *Personalized Medicine* 4(4): 447–452.

Roche, P.A., and G.J. Annas. 2006. DNA testing, banking, and genetic privacy. *New England Journal of Medicine* 355(6): 545–546.

Shabo, A. 2005. The implications of electronic health records for personalized medicine. *Personalized Medicine* 2(3): 251–258.

Shay, A. 2007. Translating from the laboratory to the patient: a multidisciplinary approach to delivering individualized therapy. *Personalized Medicine* 4(4): 435–438.

Skloot, R. 2006, April 16. Taking the least of you. *New York Times Magazine*, 38–45, 75, 79–81.

Smith, C.B., L.P. Francis, M.P. Battin, J. Botkin, J.A. Jacobson, B. Hawkins, E.P. Asplund, and G.J. Domek. 2004. Are there characteristics of infectious diseases that raise special ethical issues? *Developing World Bioethics* 4(1): 1–16.

Sobel, M.E., and S.R. Wolman. 1999. Ethical considerations in the use of human tissues in research. *Cytometry B* 38: 192–193.

Vasquez, S. 2004. Personalized therapy: an interdisciplinary challenge. *Personalized Medicine* 1(1): 127–130.

Woodward, B. 1997. Medical record confidentiality and data collection: current dilemmas. *Journal of Law, Medicine and Ethics* 25: 88–97.

World Medical Association Declaration of Helsinki. 1964. http://www.wma.net/en/30publications/10policies/b3/17c.pdf Accessed, April 3, 2013.

Zimmer, C. 2010. How microbes defend and define us. *New York Times*, July 12, Science.

Zimmer, C. 2011. Our microbiomes, ourselves. *New York Times*, December 3, Science.

1

The Human Microbiome

SCIENCE, HISTORY, AND RESEARCH

Lily Frank, Keith Benkov, Martin Blaser, Kurt Hirschhorn, Daniel A. Moros, Sean Philpott, Matthew E. Rhodes, Rhoda Sperling

Introduction

The period from 1840 to 1910 was an extraordinarily rich time for the development of modern biomedicine. Concepts, therapies, and frameworks that emerged during these decades became part of our intellectual and technological toolkit. These landmark developments include cell theory and cellular pathology, antiseptic followed by aseptic surgery, discrediting of the notion of spontaneous generation, vaccines, pharmaceuticals, evolutionary theory, genetics, and neurobiology.

During this period, the educated world also became riveted by the identification and control of bacterial organisms and launched a great battle against infectious disease. It is quite understandable that bacteria were seen as the enemy at that time. In 1840 life expectancy in the United States and United Kingdom was about 40 years; 30% to 50% of children born in the rapidly growing urban industrial centers of the West died before age five, and death was first and foremost the result of infectious disease (Barket and Drake 1982). The metaphors of battle and war were commonly used to describe the relationship between humans and microorganisms, and between 1900 and 1910 five of the first eight Nobel Prizes in physiology and medicine were given to individuals associated with a breakthrough in the battle with infectious disease.

Despite the focus on contagion and disease, as early as the 1850s Louis Pasteur and others recognized that microorganisms might play other critical roles in our environment. Researchers were concerned with the processes of putrefaction, fermentation, and, somewhat later, nitrogen fixation and were examining the role of microbes in recycling biological materials into inorganic substances. The term *biosphere* was first coined in 1875 by the Austrian geologist Eduard Seuss.[1] It is worth reminding ourselves that in science, and especially applied science, old insights often await the development of new technology and understanding before they are broadly appreciated; the process of discovery often involves rediscovery. It is only now with advances in microbial ecology and our ability to collect and identify pieces of DNA within the environment (i.e., metagenomics) that we can sort through the variety and relative prominence of the many microbial cohabitants of our bodily space. As biomedicine turned its attention to the broad interface between the microbial world and multicellular organisms such as ourselves, Nobel laureate Joshua Lederberg coined the terms *microbiome* and *human microbiome*.

A microorganism is perhaps best defined as an organism that cannot be seen with the naked eye but can be seen with a light or electron microscope. Most of these organisms are single celled and they compose small cell clusters. Microorganisms, also known as microbes, include bacteria, archaea, algae, protozoa, viruses, and some fungi.

Microbiology is the study of these diverse organisms. Bacteria and archaea are relatively simple organisms with a prokaryotic cellular structure and are usually 0.3 to 2.0 micrometers in diameter.[2] A prokaryotic cell structure is simpler than a eukaryotic one and is generally composed solely of a cell membrane surrounding a nucleoid, RNA, and cytoplasm (Smil 2002, 1). Protozoa are unicellular organisms with a eukaryotic cell structure. They include members of the genera *Plasmodium*, the causal agents of malaria, and *Giardia*, the cause of a variety of diarrheal illnesses (Cook, Zumla, and Manson 2009). A eukaryotic cell contains additional membrane-bound compartments that separate various functions (Sleigh 2006). Such structures almost always include a nucleus, where DNA is managed and selectively transcribed to RNA, and mitochondria, where food stuffs (i.e., fuels) are oxidized and energy is repackaged into the universal currency of adenosine triphosphate (ATP). Viruses are also considered microorganisms. These minute microorganisms, often not even visible with a light microscope, lack the structure and complexity of prokaryotes. Viruses do not possess the cellular machinery to reproduce on their own, and consequently, viruses, which include polio, smallpox, and human immunodeficiency virus (HIV), must infect a host cell to reproduce (Minor 2007).

Although the majority of microorganisms on this planet are viruses, bacteria generally play the most prominent role in the human microbiome. So, in what follows we shall focus primarily on bacteria. There are approximately 10^{30} bacteria on earth, with a biomass greater than all other organisms combined (Office of Science & Technology Policy and the National Science & Technology

---

[1] Because Seuss's interest was focused on geology and the Alps, he did not discuss microorganisms and marine life.

[2] There is some opposition to the use of the word *prokaryote* today, due to the fact that prokaryote is in fact a paraphyletic term (i.e., it does not refer to a single ancestral lineage).

Council, Committee on Science, Subcommittee on Biotechnology, n.d.). Colonizing the earth for roughly four billion years, bacteria have evolved to survive in an incredibly diverse range of environments, from the human intestinal tract to hydrothermal vents at the ocean's floor to the surfaces of rocks found kilometers below the earth's surface (Arumugam et al. 2011; Emerson and Moyer 2002; Maczulak 2011). Microorganisms are evolution's master survivors. Two stunning examples are the bacterium *Deinococcus radiodurans,* which can survive and repair itself after being exposed to 1,000 times more radiation than a human can tolerate, and the archaeum *Methanopyrus kandleri,* which can tolerate temperatures of up to 122°C (DeWeerdt 2002).

Microorganisms rarely live alone as single organisms or within an environment inhabited by only a single species. The recent discovery of a single-species ecosystem in a gold mine located 2.8 kilometers below the earth's surface is an uncommon exception (Chivian et al. 2008).[3] Rather, microorganisms usually exist in communities where they interact both with other microorganisms and with macroscopic organisms, such as plants and animals. Within these microecosystems, microorganisms enter into a variety of relationships with their cohabitants. These relationships may be symbiotic, in which one type of microorganism depends on another to survive and reproduce in a given environment. Microorganisms also often compete with each other for both space and resources. The very same microbes may alternate between competition and cooperation depending on the circumstances. Microorganisms often produce metabolic by-products, such as organic acids or alcohols that have an antibiotic effect, inhibiting the growth of other microorganisms. In fact, our first antibiotic drugs and many of our current antibiotic drugs are substances discovered through systematically studying the inhibition of one microorganism by another. The relationships between microorganisms and their hosts will be discussed in greater detail later in this chapter.

Traditional microbiology focuses on culture-based techniques in which microorganisms are grown in the laboratory in isolation from their host, or natural environment, and their usual microecosystem cohabitants. This approach creates a restricted, highly artificial environment, which in turn limits what can be learned about the microorganism in question. It has been estimated that less than 1% of all microorganisms can be cultured in a laboratory (Xu 2010, 1–2). Until recently, these limitations posed a major obstacle to the scientific assessment of microbial diversity. These limitations hindered our ability to understand the complex interactions among microorganisms as well as between microorganisms and their multicellular hosts. In the past few decades, new techniques in the burgeoning field of environmental

[3] A single-species bacterium has also been found in granite (Szewzyk, Szewzyk, and Stenstrom 1994).

microbiology have enabled scientists to gain far greater insight into microbial habitats. Metagenomics, the study of DNA extracted directly from an environment, in particular, has revolutionized the manner in which scientists study microbial environments. Current metagenomic techniques allow researchers to study DNA from numerous organisms at the same time, without the use of cultures.

These new techniques offer medicine and public health a broader framework for thinking about the relationship between microbial and human life.

History focuses on the search for enemy pathogens and the battles against infectious diseases such as the historically devastating specters of plague and smallpox; the continuing menace of epidemics such as HIV, influenza, and malaria; and the truly horrifying bird flu and Ebola virus. Nevertheless, the common negative connotation that has been associated with words such as *microbe* and *germ* can be misleading. Humans and most other forms of life are in fact dependent on microorganisms for their very existence. In the next section, we provide examples to illustrate the essential role that microorganisms have played and continue to play in the function of the earth's ecosystems.

The Role of Microorganisms in the Environment

Microorganisms have and continue to essentially shape the environment of the earth. In so doing, microorganisms have paved the way for the evolution of eukaryotes, multicellular organisms, and eventually humans. Without microorganisms, there would be no oxygen in the atmosphere and consequently no ozone shield; surface temperatures would be about 55°C higher than they are today, and many essential nutrients would not be readily available. Indeed, without the influence of microorganisms on the temperature and atmosphere of the Earth, it is likely that the Earth would have remained inhospitable to all but a select few microbes.

ATMOSPHERIC OXYGEN

It may seem farfetched that something as small as a bacterium, with a mass on the order of 10^{-12} grams, could affect anything as huge as Earth's atmosphere. After all, the mass of the Earth's atmosphere is approximately 10^{21} grams or 10^{33} times more massive than an average microbe. Only recently have scientists begun to fully appreciate the scope of the microbial world. It is now believed that over 90% of all microorganisms live beneath the surface of our planet. When we consider that recent calculations estimate that there are 5×10^{30} microbes on Earth, weighing a combined 5×10^{18} grams, it becomes much more plausible that microorganisms can and do impact Earth's surface on billion-year, million-year, and even shorter time scales (Whitman, Coleman, and Wiebe 1998). The number of microbial cells on the planet far exceeds those of eukaryotic organisms, and their combined biomass exceeds that of all animal life.

When considering the impact of microorganisms on Earth's surface, we have to consider the surface conditions of the primordial Earth, roughly 3.5 billion years ago. In contrast to our current atmosphere, Earth's primordial atmosphere was oxygen poor but carbon dioxide and methane rich. These gases produced an enhanced greenhouse effect, and thereby maintained warm atmospheric temperatures, more than offsetting the reduced radiant energy reaching the Earth because of the then-lesser luminosity of our young sun (Kasting 2005). Although carbon dioxide could have been provided by volcanic outgassing, methane likely was produced by anoxic methane producing microbes. The fact that there was virtually no oxygen available meant that there was no ozone shield and that the land surface was essentially barren and sterilized by intense solar radiation. Life as we know it today simply could not exist in these conditions.

Roughly 2.7 billion years ago, with the emergence of cyanobacteria, these conditions began to change (Berkner and Marshall 1965). Cyanobacteria transform energy via a process termed *photosynthesis*. During (oxygenic) photosynthesis the energy of sunlight is captured and stored within chemical bonds, and oxygen is released into the environment.

With the advent of cyanobacteria, the so-called great rise of oxygen began. It took approximately one billion years for oxygen to achieve significant levels in the atmosphere, but the effect of the rise of oxygen on life on Earth was profound and so far irreversible.

Oxygen at its current atmospheric concentration, although vital for humans, is highly toxic to many other forms of life, including the methane-producing microbes. About 2.7 billion years ago, when oxygen levels rose, the majority of ancient forms of life were either suppressed or relegated to environments where oxygen did not penetrate. It is likely that the great rise of oxygen caused the first mass extinction in Earth's history. A second consequence of the rise of oxygen was the establishment of an ozone shield. This allowed for the land surface to be colonized, first by microorganisms, then, about 400 million years ago, by plants and animals. As a consequence of this astounding environmental change, all of the familiar oxygen-related phenomena, like the burning of fire and the rusting of metal pipes, became possible—all as a result of the activity of microorganisms.

TEMPERATURE AND WEATHERING

Because of the reactivity of methane to oxygen, the great rise of oxygen also increased the rate of methane removal from Earth's atmosphere. Even as the rate of methane production by microorganisms was increasing as a result of increasingly available organic matter, both the quantity and residence time of methane in Earth's atmosphere decreased dramatically. This process effectively minimized the greenhouse warming capability of the atmosphere (Pavlov and Kasting 2002). At the same time, the sun was increasing in luminosity, thereby reducing the degree of greenhouse effect necessary to maintain temperate temperatures. The end result was an environment with a more temperate and stable atmosphere. Again, this was all the result of the activity of microorganisms.

Modern microorganisms are far more intimately involved in global temperature regulation than merely acting as methane producers. On a short-term (thousand-year) time scale, the carbon dioxide present in the atmosphere is determined by achieving equilibrium between the various carbon reservoirs at the Earth's surface. On longer, million-year time scales, the carbon dioxide level, and thus to a large extent global temperature, is determined by the flux of exchangeable carbon to and from the Earth's surface. Carbon dioxide is added to the surface of the Earth via volcanism. It is removed via the atmospheric breakdown or weathering of silicate rocks forming bicarbonate compounds. Thus, ultimately, on a million-year time scale, temperature is very much influenced by the balance between volcanism and weathering (Schwartzman 1999).

Microbes dramatically increase the rate of weathering on Earth in a number of ways. For example, microorganisms release organic acids, which aid in the breakdown of rock material; they help create microenvironments more conducive to the dissolution of rocks; and they help stabilize soils, greatly enhancing both water retention and the surface area available for erosion (Schwartzman 1999). Consequently, microorganisms are part of a multicomponent system of checks and balances for global temperature regulation. As temperatures increase, weathering rates also increase. This in turn increases the rate of drawdown of carbon dioxide, which continues until a cooler temperature is reestablished. Unfortunately, it takes on the order of a million years to establish equilibrium in this manner, and thus, natural biotically enhanced weathering does not offer a practical solution to the current climate crisis.

NITROGEN FIXATION AND NUTRIENT CYCLING

Microorganisms also play a critical role in soil formation. Often, the first organisms colonizing bare rock are either bacteria or lichen (a symbiotic association of a fungus and either an alga or a cyanobacteria) (Heritage, Evans, and Killington 1999). Once soil formation is well under way, microorganisms are responsible for much of the nutrient cycling that occurs. Foremost among these is the nitrogen cycle. All life requires nitrogen. While carbon, oxygen, and hydrogen obtained from water and atmospheric CO_2 constitute most of the Earth's biomass and constitute almost all the atoms found in carbohydrates and lipids, nitrogen is a prominent component of proteins and nucleic acids (Smil 2002). Although nitrogen gas (N_2) composes approximately 80% of our atmosphere, it is not present in a bioavailable state. A relatively small number of bacterial and algal species are capable of converting nitrogen gas to nitrogen compounds that are accessible to other forms of life via a process called nitrogen fixation. These microorganisms convert nitrogen gas (N_2) to ammonia (NH_3), which can be used by plants that they colonize through symbiotic and commensal relationships. Until the twentieth century, these microorganisms provided essentially all the available nitrogen for life on the planet.[4] Some of the nitrogen-fixing microbes have established symbiotic relationships with higher organisms, such as the leguminous plants and termites.

Microorganisms are integral components in virtually every geochemical cycle, including those involving macronutrients such as sulfur and phosphorous, micronutrients such as iron and calcium, and trace elements such uranium and arsenic (Butcher et al. 1992). It is the diversity of metabolic pathways employed by microorganisms that allow them to have such a profound impact on our planet. Microorganisms are ultimately responsible for the atmospheric O_2 and O_3 that protect our planet from cosmic radiation, stabilize (i.e., buffer) our planet's temperature, and provide habitats and nutrients for other classes of organisms. In short, microorganisms impact just about everything; animals and plants are merely along for the ride.

The Use of Microorganisms in Industry and Food Production

Throughout human history human beings have utilized the productive power of microorganisms in food and alcohol production, agriculture, industry, and, increasingly, medicine.

Humans exploit the properties of microorganisms in a multitude of ways. We use the cells themselves, such as yeast, a single-celled fungus, to make bread rise. We also extract and use the enzymes that microbes produce and their metabolic end products in food production and many other industrial, agricultural, and medical applications. Fermentation by microorganisms has been used at least since ancient Mesopotamia. Several kinds of dairy products are made with the use of acid-producing bacteria, such as *Lactobacillus bulgaricus* and *Streptococcus thermophilous*. These bacteria consume the lactose found in milk and produce lactic acid to form yogurt or cheese. Alcohol is produced through a different fermentation process. Wine and beer production involves the use of yeast, which consumes carbohydrates and produces alcohol and carbon dioxide as by-products. High-fructose corn syrup, which is ubiquitous in processed food, is made by treating cornstarch with several enzymes that are produced by bacteria (Thomas 2004).

---

[4] With the development of the Haber-Bosch process, massive amounts of ammonia (fixed nitrogen) are now produced by the chemical industry. Without these products, modern agriculture would not be possible.

Bacteria such as *Rhizobium* sp. are being tested for use in biofertilizers because of their ability to convert the nitrogen plentiful in our environment into a form that plants can utilize (Alami et al. 2000; Fuentes-Ramirez and Caballero-Mellado 2005). Bacteria such as *Bacillus thuringiensis* that thrive on plants can also be used in agricultural applications because of their ability to produce a protein that acts as an insecticide against crop-consuming caterpillars and moths (Krimsky and Wrubel 1996). Using bacterial rather than chemical insecticides is preferable because it minimizes the impact on the environment, water, and human health (Krimsky and Wrubel 1996).

Compounds that microorganisms produce are also used in industrial applications. Bacteria that can thrive at high temperatures, called thermophilic bacteria, are used to age denim in textile factories. Enzymes produced by thermophilic bacteria are used in laundry detergent (Facklam & Facklam 2004). In the emerging field of biomining, *Thiobacillus ferrooxidans*, for example, is used in copper and gold mining because it consumes the inorganic materials found in the metal/ore amalgamation and releases acids, which through oxidation separates the metal from the ore (Thomas 2004).

Microorganisms also can be used to neutralize toxic industrial wastes through a process called bioremediation in which bacteria feed on oil and gasoline spills. Once their food source is used up, the microorganisms die (Environmental Protection Agency, n.d.). Along similar lines, the bacterium *Pseudomonas stutzeri* is used in repairing damaged works of art. In 2003, researchers applied the bacteria to fourteenth-century Italian frescoes that had been damaged by the use of glues and cleaning agents on their surface. The bacteria consumed the damaging agents and revealed the original work (Thomas 2004).

The History of Microorganisms in Human Health and Disease

To illustrate the evolving scientific understanding of the complex relationships, harmful and beneficial, between microorganisms and human beings, we offer a brief outline of the discovery of the role that microbes play in disease.

Microorganisms first became visible to humans when Anton van Leeuwenhoek (1632–1723) observed protozoa in rainwater through his microscope in 1674. Later, in 1683, he observed what he called "animalcules" and "wretched beasties" that he found in the material he scraped off of his teeth (Tannock 1994). Van Leeuwenhoek famously wrote, "all the people living in our United Netherlands, are not as many as the little animals that I carry in my mouth this very day" (Dobell 1932). The discovery of these "animalcules" was soon followed by an "animalcular" theory of infectious disease. Because there was no evidence at the time that these "beasties" actually caused the spread of disease, this idea amounted to a hypothesis.

Shortly thereafter, in the early 1700s, Cotton Mather in Boston, a correspondent of the Royal Society and a strong supporter of smallpox inoculation, wrote of the "animalcular" nature of smallpox (Hopkins 2002, 10). Almost a century later, Edward Jenner (1749–1823), routinely referred to as an "English country doctor" but also a member of the Royal Society, developed a relatively benign and effective vaccination against smallpox. At the time, smallpox produced greater mortality than any other infectious disease in Europe or North America (Hopkins 2002, 32). Based on the well-known fact that survival from an attack of smallpox produced lifelong immunity, inoculation with the fluid of smallpox vesicles (with the objective of producing a mild case of the disease) had become a fairly common practice in England and America. Jenner observed that milkmaids, who were exposed to cowpox, then a fairly uncommon disease of cattle, were not infected with smallpox, nor would they get a mild form of the disease when inoculated with material from smallpox vesicles. After extensive investigation of both these diseases, Jenner began using material from vesicles in cowpox rather than smallpox, producing immunity with much less morbidity and mortality. Jenner published these findings in 1798 (Pommerville 2011, 14). Indeed, the cowpox vaccination offers the first example of the explosive spread of a medical technology in early modern/modern Europe. By 1800, Thomas Jefferson was obtaining material for use at Monticello (Hopkins 2002, 264–265), and by 1807, the first compulsory vaccination of schoolchildren was introduced in Bavaria (Baldwin 2005). Indeed, in 1805, with the Napoleonic Wars raging, the emperor is reported to have said to Josephine when considering a petition from Jenner for the release of two persons, "Jenner, ah, we can refuse nothing to that man" (Saunders 1982, 164).

It has been recognized since the beginning of "scientific microbiology" in the midnineteenth century that microbes both threaten us through disease and contribute to our ability to maintain health. In the late 1850s, before he developed the "germ theory" of disease, French microbiologist and chemist Louis Pasteur (1822–1895) was coming to the surprising conclusion that the ferments associated with the formation of alcohol, lactic acid and butyric acid, were actually living organisms. By the early 1860s he already coined the terms *aerobic* and *anaerobic* and concluded on "the basis of limited observations, and without extensive experimental evidence…that similar phenomena occur during putrefaction and that the evil smelling decomposition of beef bouillon, egg albumin, or meat is the result of anaerobic life of specialized germs that attack proteins under the protection of aerobic forms capable of removing oxygen from the environment" (Dubos 1960, 116–157, especially 134–137).

In addition to this synergistic interaction between microorganisms, bacterial antagonism was also noted. Indeed, in 1876, in a letter to biologist Thomas Huxley (1825–1895), John Tyndall (1820–1893), a prominent experimental physicist, best remembered today for the "Tyndall effect," wrote that "the most extraordinary cases of fighting and conquering between the bacteria and the penicillium have been revealed to me" (Kingston 2004, 444). This letter is quoted at greater length by Ronald Clark in his biography of Ernst Chain (1906–1979), a German biochemist who shared the Nobel Prize with Alexander Fleming and Howard Florey for his work with penicillin:

> ... mutton in the study gathered over it a thick blanket of Penicillium. On the 13th it had assumed a light brown colour, 'as if by a faint admixture of clay'; but the infusion became transparent. The clay here was the slime of dead or dormant Bacteria, the cause of their quiescence being the blanket of Penicillium. I found no active life in this tube, while all the others swarmed with Bacteria.
>
> In every case where the mould was thick and coherent the Bacteria died, or became dormant, and fell to the bottom of the sediment. The growth of mould and its effect on the Bacteria was very capricious.... The Bacteria which manufacture a green pigment appear to be uniformly victorious in their fight with the Penicillium.... (Clark 1985, 26)

Tyndall "went on to identify three different strains of penicillium, distinguishing them by their cultural characteristics, but he does not appear to have realized that inhibition of the bacteria was due to a chemical substance produced by the mould" (Clark 1985, 26). In the following year, 1877, Pasteur wrote:

> In the inferior and vegetable species, life hinders life. A liquid invaded by an organized ferment, or by an aerobe, makes it difficult for an inferior organism to multiply.... These facts may, perhaps, justify the greatest hope from a therapeutic point of view.

This remarkably sophisticated set of questions arose very early in the history of microbiology. By the late 1870s, scientists had discovered that when bacteria that are pathogenic for humans and animals find their way to the soil, either in the excreta of the hosts or in their remains, they do not survive there for long. Bacterial antagonism had been suggested as the reason, and indeed, as early as 1881, German chemist Max Joseph von Pettenkofer (1818–1910) proposed that useful bacteria might be cultivated and domesticated in the soil under a house so that they could starve out or destroy pathogens (Kingston 2004, 446). This is a sample of the long history of researchers and physicians uncovering both the complex relationships that exist between microorganisms and the ways in which microorganisms and their products can be used therapeutically.

Pasteur went on to help develop the germ theory of disease by observing the role yeast played in fermenting wine and the role bacteria played in souring wine and spoiling milk. A chemist by training, Pasteur had been commissioned by French agricultural agencies in the 1850s to investigate diseases that affected the production of wine. During these studies, Pasteur began to champion the idea that fermentation, and by analogy putrefaction, was a biological rather than a chemical process. He claimed that it was due to the action of microorganisms or living germs, rather than molecular catalysts or chemical ferments. Ultimately, Pasteur played a crucial role in discrediting the theory of spontaneous generation and establishing the germ theory of disease. He demonstrated that bacteria are omnipresent in our environment, including the air, and that contact with these bacteria contributes to infection with disease. Pasteur also developed vaccines, including a vaccine for rabies. Not all of Pasteur's work, however, focused on the harmful consequences of microbial infection. Pasteur claimed that without microbes, the human body could not function (Falk et al. 1998). He and his assistant Jules Francois Joubert (1804–1907) were among the first to observe "interbacterial inhibition" in their study of anthrax bacillus (Riley and Chavan 2007, 46). They noted that when they attempted to cultivate anthrax along with another species of bacteria, the anthrax was unable to thrive (Riley and Chavan 2007, 46).

The notion that unseen organisms can be the agents of disease was further advanced in 1854, when physician John Snow studied the daily routines of the victims of a London cholera outbreak. He traced the source of their infection to the well that was used for drinking water. To prevent residents from drinking the contaminated water, he removed the well pump handle (Riley and Chavan 2007, 16).

During the same period, the English surgeon Joseph Lister (1827–1912), influenced by Pasteur's suggestion that there are microorganisms in the air, postulated that keeping wounds sterile with surgical dressings and cleaning surgical instruments with carbolic acid would reduce postsurgical infections (Lister 1867; Ellis 2001, 93). He also encouraged surgeons to wash their hands with carbolic acid solution before performing surgery. At the time, postsurgical infection was a very common cause of death among surgical patients (Thomson 1963, 711).

German physician Robert Koch (1843–1910) pioneered an approach to determine that a particular microorganism is responsible for a specific disease. Through his experiments, Koch discovered the bacteria responsible for anthrax, cholera, and tuberculosis, among other diseases. He also established that tuberculosis is contagious between humans. Koch developed techniques in microscopy, methods of cultivating microorganisms in the laboratory, and staining and vivisection techniques. Among Koch's contributions to microbiology was the development of widely accepted criteria (commonly referred to as Koch's postulates) for determining whether there is a causal link between a

---

[5] Koch's postulates:

1. The same microorganisms are present in every case of the disease.

2. The microorganisms are isolated from the tissue of a dead animal and a pure culture is prepared.

3. Microorganisms from a pure culture are inoculated into a healthy, susceptible animal. The disease is reproduced.

4. The identical microorganisms are isolated and recultivated from the tissue specimens of the experimental animal. (Pommerville 2011, 18)

particular bacteria and a particular disease.[5] Koch's postulates added structure to the fields of microbiology and epidemiology.

The notion that diet and certain foods may contribute to health and disease certainly extends back to the Hippocratic corpus. Very early in the development of microbiology, the salutary effects of microorganisms were noted and incorporated into the nutritional perspective as "probiotics." Probiotics are "live microorganisms which when administered in adequate amounts [are supposed to] confer a health benefit on the host" (U.S. Food and Agriculture Administration and World Health Organization 2002). During the late nineteenth century and early twentieth century, Russian scientist Ilya Metchnikoff noticed unusual longevity among populations that consumed large amounts of sour milk as part of their diet. Metchnikoff was convinced that through the cultivation of lactobacilli, which he called "Bulgarian bacillus," in the large intestine, a beneficial change in pH could be brought about that would inhibit the growth of other harmful bacteria (De Kruif 1926, 226). This early attempt at probiotic therapy was unsuccessful.

Scottish physician and scientist Alexander Fleming's (1881–1955) 1928 discovery profoundly changed our relationship to bacteria. After growing the bacteria *Staphylococcus* on a culture plate, the plate was accidentally contaminated with mold. After leaving the plate unattended for several days, Fleming observed that bacteria did not grow in the area near where the mold grew. Further experiments revealed that the mold, *Penicillium notatum,* released a substance that killed bacteria, which he called penicillin. Fleming did not specifically identify a chemical substance but rather created a liquid extract presumed to contain a specific substance.

This rich history invites wonder as to why it took so long for antibiotics to be developed, and the story of these biologic and pharmaceutical products is routinely misrepresented in most brief didactic presentations. Although the idea of "life hindering life" was well recognized before Fleming refocused attention on the phenomenon, there were technical difficulties in isolating and producing the biologic products of bacteria and fungi. At the same time, biochemistry was a young science and routine methods for separating different biochemical products did not exist. Fleming's recognition of the antibiotic product from penicillin in 1928 did not lead to a useful therapeutic product until the work at Oxford of the Austrian biochemist Ernst Chain and Australian pathologist and pharmacologist Howard Florey (1898–1968) over 12 years later.

Even after being certain of the efficacy of penicillin in the early 1940s, more than half a decade was spent developing the necessary technology for large-scale bacterial fermentation to provide adequate supplies for routine use.[6]

---

[6] See Clark (1985, pp. 70–74) for more details.

Already, in 1940, the adaptation of microorganisms to their environment was noted with the appearance and spread of penicillin-resistant *Staphylococcus* within hospitals where the drug was first used. By late 1940, biochemists Edward P. Abraham (1913–1999) and Chain, who worked extensively on identifying the structure of penicillin and how to isolate it from other substances, were reporting on a penicillin-destroying enzyme, which they named penicillinase (Abraham and Chain 1940).

When Selman Waksman (1888–1973), biochemist and microbiologist, identified the source organism for the second major antibiotic, streptomycin, in the early 1940s (working at Rutgers University with grants from Merck). It required 50 Merck scientists working for two years to complete the pilot production of a usable pharmaceutical material (Kingston 2004, 460).

The idea of isolating usable antibiotic products from microorganisms was very much "in the air" during the early 1900s. In 1927, two years before the announcement of Fleming's discovery, physician-researcher Oswald Avery (1877–1955), working at Rockefeller Institute on *Pneumococcus*, recruited French microbiologist Rene Dubos (1901–1982) to search for soil organisms that would destroy the polysaccharide capsule of type III pneumococcal bacteria (Van Epps 2006, 259). It was Avery who, working on *Pneumococcus*, discovered in the 1940s that DNA and not protein was the material that transmitted genetic information.

The Human Microbiome

As a result of the Human Microbiome Project (HMP), a broader vision of the interconnectedness of the biological world is being incorporated into medical thought (Ackerman 2012). Although researchers and scientifically minded physicians have long been aware of the interconnectedness of human beings and the microbiological world, the general public is only now becoming aware of it. This vision emerged remarkably rapidly with advances in cell biology and microbiology in the middle of the nineteenth century. Medicine is an applied science, and it applied the new toolkit of biomedical science to new practical roles within medicine.

In this section we introduce some terminology for defining the diverse types of relationships that human beings have to microorganisms. The chapter then focuses on describing the human microbiome, how it develops, and how it relates to health and disease.

Symbiosis is a general term that captures the range of relationships that can exist between organisms. Symbiosis exists any time two or more different species of organisms live in close relation to each other, usually on or in each other. Such associations include a variety of more specific relationships: parasitism, commensalism, and mutualism. Human–microbe interactions run the gamut.

Parasitism, or the symbiotic relationship where one partner benefits and the other is harmed, is the type of relationship that we most commonly think of when we consider bacteria and viruses. When microorganisms colonize a human host and develop a parasitic relationship with that host, the microorganisms are often the cause of disease. Pathogen is a related term, used in a medical context to describe microorganisms responsible for disease, but in a broadly biological context a pathogen is any microorganism that both can resist host defenses that would eliminate most other microorganisms and has the ability to cross anatomical barriers (Blaser and Falkow 2009, 888).

For example, a parasitic relationship exists between humans and the pathogen *Yersinia pestis*, the bacterium that causes the plague. These bacteria live in fleas that parasitize rodents. Humans and other animals are infected with *Y. pestis* when they are bitten by fleas and the bacteria are transmitted to them. Infection by plague bacteria now can be treated with a course of antibiotics. Benign and useful relationships between microbes and humans also exist. In fact, they make up the majority of the relationships that humans have with microorganisms (Blaser and Falkow 2009, 887). Commensalism is a type of relationship that exists between a microorganism and its host in which one of the organisms benefits from the relationship, while the second organism is neither harmed nor benefited by it. For example, *Candida albicans* is a yeast that colonizes approximately 80% of the human population. Unless the host is immunocompromised, colonization has no known deleterious effects (Blaser 1997, 760).

A mutualistic relationship between a microorganism and its host is one in which both organisms are benefited by their interaction. Mutualistic relationships between human beings and bacteria abound. For example, in the human intestine bacteria play a significant role in extracting nutrients from the food that we consume and in producing needed materials, such as vitamin K (Brooks et al., 2013). Through a wide variety of symbiotic relationships, humans and microorganisms have evolved together. We have evolved to rely on many of the processes and outputs of the genes of the microorganisms. Microorganisms have evolved to inhabit the environments that our bodies provide (Lederberg 2003, 53–55).

In many cases, the microorganisms that inhabit our bodies can play either parasitic, commensal, or mutualistic roles, "depending on the context" (Blaser 2006, 115). Examples of this phenomenon can be found throughout the human body. For example, pneumococcus and meningococcus bacteria colonize the nose and throat of many people without causing any ill effects; however, these same bacteria are responsible for illness (Falkow 2006). Martin Blaser argues that the *Helicobacter* species that colonize our stomachs as well as the stomachs of many other mammals are a perfect example of this type of relationship, called amphibiosis, a term coined by Theodore Rosebury (Blaser 2006, 115). *H. pylori*, the human gastric organism, Blaser suggests, is ancient;

it has inhabited the stomach of humans and our ancestors since at least the Paleolithic era (Blaser, Chen, and Reibman 2008). Both the beneficial and harmful consequences of colonization with *H. pylori* have been widely studied. For example, persistent *H. pylori* infection plays a role in the incidence of gastric ulcers and cancer; at the same time, Blaser has argued that the decreasing presence of *H. pylori* infection in developed countries may be causally linked with a rise in the rate of esophageal cancer, asthma, and allergies (Blaser and Atherton 2004; Blaser et al. 2008).

A large body of research in a variety of scientific disciplines shows that the relationship between microorganisms and their human hosts is far more complex than a simple adversarial model. Pathogenic or disease-causing microorganisms constitute only a small portion of the trillions of microorganisms that colonize the human body. The non-disease-causing microorganisms have either a commensal or a mutualistic relationship with their hosts, meaning that both organisms benefit.

In and on our bodies, microbial cells outnumber human cells by a factor of 10 to 1. The number of bacteria occupying the human body is estimated to be in the trillions (Curtin 2009), and it is estimated that there are two to four million distinct nonhuman genes in the human body (Feldman 2004, 205). Microorganisms, especially bacteria, populate many areas of the body, with the intestine containing more microorganisms than all of our other colonized sites combined. We can also find substantial populations of bacteria on the skin, in the urogenital tract, and in the mucous membranes surrounding the eyes, mouth, nose, and ears (Frank and Pace 2008). Some areas of the body such as the central nervous system, blood, parts of the lungs, liver, spleen, kidneys, and bladder usually remain sterile (no living microbes are present) when healthy (Levinson, n.d.; Niven, n.d.).

The term *microbiome* is a combination of "micro," from microorganism, and "biome." A biome is "a major community of plants and animals having similar life forms or morphological features and existing under similar environmental conditions... [and] may contain several different types of ecosystems" (Risser, n.d.).

The term *human microbiome* refers to the population of microorganisms (pathogenic, commensal, and mutualistic), bacteria, viruses, fungi, and protozoans and their genetic material that live on and inside the human organism (Honey 2008; Lederberg 2003). A related term is the *human metagenome*, which is the amalgamation of both our human genes and the genes of the microorganisms that populate the inside and outside of our bodies (National Research Council 2007). Some scientists, including Nobel Prize recipient, molecular biologist Joshua Lederberg, suggest that rather than conceiving of the human being as the collection of human cells and genetic material that compose the body, we should conceive of human beings as superorganisms or human–microbe hybrids, "a community that adds up to more than the sum of its parts" (Mullard and Mandavilli 2008, 578). The health of this superorganism is determined by both "intrinsic properties such as human genetics, sex, diurnal cycles, and age and by extrinsic factors such as lifestyle choices (food and drink, drug intake) and the acquisition of a stable healthy microflora (the so-called microbiome)" (Goodacre 2007, p. 259S).

Some factors that can influence the composition of the microbiome are obvious, such as taking a course of antibiotics, designed to kill a specific type of microorganism. Other factors may be surprising. Recent studies show that in mice, one day of consuming "junk food," foods high in fat and sugar, has a profound effect on the bacteria that inhabit their intestinal tract (Turnbaugh, Bäckhed et al. 2008). In another finding, a person's sex is in part correlated with the kinds of microorganisms and the level of diversity of microorganisms that inhabit the surfaces of their hands (Fierer et al. 2008). Researchers point out that differences in microorganisms colonizing palm skin in men and women could stem from a variety of factors including differences in pH, the production of sweat and sebum, how often cosmetics were applied, hormone levels, and thickness of skin (Fierer et al. 2008).

It is possible to conceptualize the human microbiome in several ways. One prominent model conceives of the human microbiome as a multicelled organ. The functions of healthy microbiota mimic a well-functioning organ: it "consumes, stores, and redistributes energy and mediates important chemical transformations that benefit the host" (Fierer et al. 2008 2). For example, bifidobacteria, which inhabit the human intestinal tract, play a role in the maintenance of and communication across the mucosal barrier of the intestinal epithelium. More specifically, *Bifidobacterium longum* plays a role in breaking down complex molecules that come from the food we eat. For example, when we consume starches in the form of potatoes that contain polysaccharides, bacteria feed on those molecules and break them down into the simpler oligosaccharides. Bifidobacterium then further break down these molecules through a process of fermentation, producing the small molecules acetate and lactate (Kinross et al. 2008). These smaller molecules are then transformed by yet another set of bacteria into butyrate, a short-chain fatty acid that serves as a source of energy for the mucus of the colon (Kinross et al. 2008). Through "feedback loops" the microbiota and host cells are able to repair and expand appropriately, just as organs do (Foxman et al. 2008). Thinking of the human microbiome as a multicelled organ focuses attention on organ fluxes and their correlated states of disease or health.

A competing way of conceiving of the human microbiome is as a collection of diverse ecological communities that interact with each other, the host, and the environment (Foxman et al. 2008, 3). Thinking of the human microbiome in this way emphasizes the "interactions between component organisms and their dynamics" (Foxman et al. 2008, 3). The effects on the host, whether parasitic, commensal, or mutual, are, in part, the result of the interactions between microbial communities. For example, microorganisms that inhabit the vagina after puberty, such as lactobacilli, consume carbohydrates and release acid. In this way, they play an important role in creating an environment with an acidic pH in the vagina. An acidic environment is inhospitable to a variety of other microorganisms, for example, yeasts, that would otherwise have the opportunity to dominate the ecosystem. Conceiving of the microbiome as a series of interrelated ecological communities is one way to explain the movement of microorganisms from one area of colonization on the body to another. Until we know more about the ways in which the microbiome functions, it makes sense to keep both of these models in mind and to reevaluate their appropriateness in specific cases and in the light of new information (Foxman et al. 2008).

Studies performed on germ-free animals were some of the first to show that the group of microorganisms that normally inhabits an organism's body can have a significant impact on the health of the hosts. Studies on mice showed that mice raised in a sterile (i.e., microbe-free) environment had compromised immune systems, an inability to efficiently digest food, and organs that were smaller than normal (Mullard and Mandavilli 2008, 578). Although human microbiomes are not identical to those of mice, mammals that have similar diets have microbiomes that are closer to each other than mammals with markedly different diets (National Institutes of Health, n.d.-a). The roles of the normally nonpathogenic microorganisms that inhabit our bodies are varied and complex. For example, within the intestinal tract, bacteria play the essential role of synthesizing vitamin K and allowing for utilization of nutrients in the foods we eat (Ouwehand and Vaughan 2006, 141). The presence of a specific set of nonpathogenic normal microbiota on the skin and mucous membranes may protect the host from colonization. These pathogenic bacteria do this by producing substances that are toxic to the pathogenic species or effectively compete with them for food sources (Grice and Segre 2011).

Scientists have already identified a wide range of symbiotic and commensal relationships between humans and microorganisms. Now they are beginning to uncover the complex and multifaceted ways in which the microbial communities that colonize human beings impact on health and disease.

Human Microbiome Project and the National Institutes of Health

In 2007, the National Institutes of Health (NIH) launched the Human Microbiome Project. The mission of this research initiative was to characterize the microbial communities that inhabit the human body and explore the relationships between the microbiota and their human hosts, including effects on human health and disease, development, physiology, immunity, and nutrition (National Institutes of Health, n.d.-b). Some have called the HMP the second human genome project, in part because it aims to take advantage of new and emerging technologies in genetics and DNA sequencing, particularly those that enable scientists to study microorganisms that have never successfully been studied before (National Institutes of Health). In addition, the project seeks to understand microbial genetic makeup, gene expression patterns, and metabolic physiologies as they relate to microbial interspecies interactions and microbe–host interactions (National Institutes of Health). NIH invested $140 million in this five-year project (Curtin 2009).

The primary goals of the HMP are to:

1. Determine whether individuals share a core human microbiome;
2. Understand whether changes in the human microbiome can be correlated with changes in human health;
3. Develop the new technological and bioinformatic tools needed to support these goals;
4. Address the ethical, legal, and social implications raised by human microbiome research;
5. Engage in international collaboration to generate a rich, comprehensive, and publicly available data set (Curtin 2009.). This fifth objective includes establishing the Human Microbiome Project Data Analysis and Coordination Center to serve as a data warehouse for all of the information that the HMP produces and later contribute to other public databases (Honey 2008).

At the same time a related international effort is under way, including similar projects in Canada, France, China, Japan, Singapore, and Australia. The second-largest project after the United States' HMP is the European Commission's Metagenomics of the Human Intestinal Tract (MetaHIT), which will study the microbiota of the human intestinal tract and their relationship to obesity and inflammatory bowel disease (Mullard and Mandavilli 2008, 578). The International Human Microbiome Consortium seeks to bring researchers from related efforts together for data sharing.

Both the HMP and MetaHIT aim to sequence bacterial genomes and to make these data available to other researchers through a database. The HMP intends to sequence more than 1,000 bacterial genomes and MetaHIT more than 100 to use as reference genomes. Reference databases will enable scientists to compare the functions and characters of genes that have been sequenced and studied with genes from microbes that are presently not well understood or difficult to culture. Organisms that contain genes similar to those that already have been entered in the database may be expected to have similar metabolic pathways (Mullard and Mandavilli 2008, 580).

Research Tools and Methods

The classic view of a microbiologist is a scientist armed with a microscope and some Petri dishes. While the tools of a microbiologist have certainly come a long way since Leeuwenhoek's first microscope, but in some sense little has changed. Both microscopy- and culture-based techniques remain widely used. It is certainly beyond the scope of this book to discuss all scientific techniques relevant to the human microbiome. Nevertheless, it is worthwhile to briefly review some of the methods most commonly employed to investigate microbial communities and the human microbiome.

MICROSCOPY

By simply counting, staining, and analyzing cell morphologies, researchers can quickly acquire a rudimentary knowledge of an environment's population density and diversity. More modern hybridization-based techniques, such as fluorescent in situ hybridization (FISH), can offer much greater insight into an environmental microbial community. In FISH, short sequences of DNA tagged with fluorescent molecules, termed *probes*, are mixed with a sample. The probes then bond to regions of DNA with highly similar sequences and fluoresce, allowing scientists to identify specific categories of cells. FISH probes can be designed to target a broad array of microbes, ranging from individual species to essentially the entire microbial world.

Often a number of different probes, each fluorescing a different color, are employed together to identify multiple categories of cells at the same time. However, the power of FISH and similar methods lies in the ability to combine them with other complementary techniques. These days the molecular and atomic compositions of individual cells can be ascertained with great accuracy. By combining the identifying capacity of FISH with molecular analyses of cells, researchers can ascertain the roles and metabolisms of various classes of microbes in an environment.

HIGH-THROUGHPUT METHODS

While microscope-based methods are invaluable for the detailed analysis of relatively small numbers of microbes, often researchers are more interested in the larger environmental picture. Two techniques commonly employed to investigate the bulk properties of microbial communities are denaturing gradient gel electrophoresis (DGGE) and microarrays (Fischer and Lerman 1979). In DGGE and other related electrophoresis techniques, the negative charge inherent to DNA or RNA, or in the case of protein, an imparted negative charge, is utilized to drive the molecule under investigation through a porous medium or gel. This feat is accomplished by running an electric current through the gel. The larger the molecule, the more difficult it is for it to move through the gel and the slower its progress; therefore, DNA, RNA, and proteins can be separated by size.

In the case of DGGE conducted on a DNA sample, generally the environmental diversity of a single gene is under investigation. The gene of interest is first amplified to produce a sufficient quantity for analysis. The DNA is then run through a gel and a denaturing chemical that causes the DNA to unfold is applied in a gradient to the gel. The denaturing chemical causes each individual DNA molecule to spread apart and essentially grow in size at a unique point in the gel. Sequence differences down to individual nucleotides can be detected in this manner.

Thus, scientists can learn a great deal about the diversity of an environment with respect to a single gene without determining the sequences of the DNA molecules in question.

In contrast, microarrays are highly efficient tools for analyzing the broader diversity and metabolic activity of an environment, once more without resorting to sequencing the individual microbes. In microarrays, a small chip of glass is spotted with thousands of small molecules (again termed *probes*) that bond to specific DNA, RNA, or protein targets. The probes themselves generally consist of molecules of DNA, RNA, or antibodies. Then a collection of fluorescently labeled sample DNA, RNA, or proteins is washed over the microarray and hybridized to the probes. By analyzing the spots on the microarray that fluoresce, researchers can determine which nucleotide sequences or proteins are present in an environment. In this manner, scientists can relatively quickly ascertain both the overall diversity of an environment and the metabolic processes conducted therein.

METAGENOMICS

Since its advent in the 1970s (Maxam and Gilbert 1977; Sanger and Coulson 1975), DNA sequencing has become an indispensible part of microbiological studies. Currently, the vast majority of environmental microbiological studies rely at least partly on sequence data. Traditional microbial genetic studies focused on culturing individual species of bacteria in laboratory settings. Such approaches have two substantial deficiencies. First, culture-based studies are far removed from the natural environment and offer little insight into the interactions between microorganisms. Second, although feasibility in this area is improving, it has been estimated that less than 1% of the bacterial community found in most environmental samples can be cultured in a laboratory, thereby greatly restricting the scope of culture-based studies (Amann, Ludwig, and Schleifer 1995).

To overcome these limitations, the burgeoning field of metagenomics or environmental genomics was born. Metagenomics refers to the study of genetic material extracted directly from an environmental sample. The term was coined by bacteriologist Jo Handelsman in the course of her work on soil microbiology (Handelsman et al. 1998). The first metagenomic studies in the late 1980s relied on cloning of environmental DNA (Lane et al. 1985). In cloning, DNA from an environmental sample is stably inserted into a host cell. The host cell is then grown in culture, producing copies of the inserted DNA for traditional sequencing. Similar cloning techniques are still commonly used today.

The advent of modern next-generation sequencing technologies has made large-scale metagenomic studies possible. Next-generation sequencing technologies offer three distinct advantages over traditional sequencing methods. They are much cheaper than traditional sequencing technologies; they produce thousands, if not millions, of times more sequence data than traditional sequencing technologies; and they generally do not require a cloning step. The drawbacks of next-generation technologies are twofold. First, the read length, or the amount of DNA sequenced per sequencing reaction, is shorter, making it harder both to identify the origin of the DNA and to assemble the individual reads into longer segments. Second, next-generation sequencing technologies are often more prone to sequencing errors. Nevertheless, with increasingly powerful computational tools and ever-improving technologies, these obstacles can be surmounted. For example, machines now are available that are substantially lengthening the short reads previously obtained with very-high-throughput sequencing.

Once DNA has been extracted from an environmental sample, a metagenomic study can proceed by using either 16S rRNA and other individual genes or random sequencing of environmental DNA. Both of these methods will be discussed in the following sections.

16S RRNA AND OTHER INDIVIDUAL GENE-BASED STUDIES

The 16S rRNA gene encodes an essential component of the bacterial ribosome. The 16S rRNA gene is universal in prokaryotic life and is one of the most highly conserved elements in all prokaryotes. The implication is that 16S rRNA function is highly important to cells and changes in the 16S sequence almost always result in negative consequences. Therefore, 16S sequences are highly similar across species, even when these species evolved in separate environments or when their evolution was separated by millions of years. Eukaryotic organisms encode a similar, though slightly larger, ribosomal component. The universality and conserved nature of the 16S rRNA gene make it an ideal choice for characterizing the diversity among microbes. It was Carl Woese in the late 1970s who first attempted to use these unique attributes of the 16s rRNA gene to investigate the phylogenetic relationships between different lineages of bacteria (Woese and Fox 1977). Later, Woese's methods were adapted to study environmental samples. By looking at the diversity and distribution of 16S rRNA sequences in an environment, researchers can quickly and easily ascertain the diversity and distribution of microorganisms in that environment. Essentially, the 16S rRNA sequences serve as a proxy for the entire genome.

The first environmental microbial surveys utilizing the 16S rRNA gene were conducted by Norman Pace in the mid-1980s (Lane et al. 1985). As traditional sequencing technologies require copious quantities of identical fragments of DNA, obtaining DNA of sufficient quantity and quality proved to be quite difficult. The method Pace created evolved into the now-widely used procedure of (1) extracting environmental DNA; (2) using polymerase chain reaction (PCR) to produce numerous copies of (i.e., amplify) the 16S rRNA sequence; (3) inserting a single copy of the PCR amplicon into a host microorganism, typically *Escherichia coli*; (4) growing the host microorganism to produce numerous copies of the inserted sequence; (5) extracting DNA from the host organism including the inserted sequence; (6) reamplifying the inserted 16S rRNA sequence; and finally (7) ascertaining the exact 16S rRNA sequence. The procedure is labor intensive and produces a single 16S rRNA sequence per host cell. Other genes restricted to specific lineages of bacteria or specific metabolic pathways are also occasionally targeted when more focused questions are addressed.

Once again, the advent of next-generation sequencing technologies has revolutionized the field. Due to cost and labor demands, before the advent of next-generation sequencing technologies, 16S rRNA gene libraries generally consisted of tens to hundreds of sequences. Rarely did a library consist of a thousand sequences (i.e., copies of a gene). Next-generation sequencing technology allows for portions of the 16S rRNA gene to be sequenced tens of thousands of times or more, thereby enabling a much more detailed look at the diversity of an environment without some of the biases inherent to the traditional method. This allowed Mitch Sogin in 2006 to propose the existence of a "rare biosphere," or a community of microorganisms each composing one millionth or less of the total microbiome (Sogin et al. 2006). Despite their rarity, the microorganisms making up the rare biosphere may constitute the majority of the diversity in an environment and may play a critical role in both environmental processes and evolution.

RANDOM SEQUENCING OF ENVIRONMENTAL DNA

Many of the metagenomic studies published today focus on random sequencing of environmental DNA from highly diverse environments. Sequencing of this nature provides a cross-section of an environment's microbiome. The kinds of organisms and metabolic pathways that are most common in the sample are represented by the largest number of sequences. The process of random sequencing involves first cutting the environmental DNA into small pieces of the appropriate size for the sequencing technology to be applied. The DNA is then sequenced and, when possible, the reads are assembled into larger fragments. The degree of assembly is dependent on the complexity of the environment, the quantity of DNA sequenced, and the available database relevant to that site. Even with little to no assembly, valuable information can be obtained from a metagenomic data set. The first studies using this technique demonstrated the unexpectedly high diversity among the microbial communities in a marine viral community (Breitbart et al. 2002) and in a Sargasso Sea microbial community (Venter et al. 2004). More recent studies have focused on the overall metabolic capabilities of particular microbial communities (Tringe et al. 2005) and the distribution of genes among different environments (Dinsdale et al. 2008).

The techniques and analysis tools of metagenomics are evolving rapidly. One challenge facing metagenomicists today is the difficulty in extracting meaningful results from the ever-expanding quantity of sequence data produced by modern DNA sequencers. For example, the occurrence of an organism's DNA in an environmental sample does not necessarily indicate that the organism is indigenous to the environment. Contamination of samples can occur during laboratory procedures or in nature, as when sewage flows into the ocean. Nor does the occurrence of an organism in an environment necessarily indicate that the organism is actually functioning in the environment. Many microbial species, including some members of the phylum *Firmicutes*, are capable of entering a stasis state during unfavorable periods, often to avoid desiccation or nutrient deprivation. In such cases, although their DNA is present in the environment, the organisms themselves are not actually metabolizing or contributing to the environment. Although an organism may contain a gene for a specific function, the gene itself may only be active under specific environmental conditions. For instance, if an environment is rich in a particular amino acid, a microbe that is capable of producing the amino acid by itself may refrain from doing so.

Thus, while metagenomics has certainly become the gold standard in microbiome studies, it is important to recognize that metagenomics has its limitations and can often present a biased view of an environment. Consequently, metagenomic studies should be complemented by other techniques, including, but not limited to, previously mentioned.

Acquisition of the Microbiome

One of the central questions that researchers studying the human microbiome are attempting to answer is how and when the microbiome is acquired. The time that a human being spends in its mother's womb is the only period in a normal life in which the body is relatively free of microbes. An individual's microbiome is acquired after birth from environmental sources. (Dominguez-Bello et al. 2010). Chana Palmer and colleagues, researchers at Stanford University, note that within the first year of life, the microbiota of the gastrointestinal tract transforms from being virtually nonexistent to "an extremely dense colonization, ending with a mixture of microbes that is broadly very similar to that found in the adult intestine" (Palmer et al. 2007, 1557). The process by which this occurs is not well understood.

Although the acquisition of the gut microbiome in infants has been studied, a clear body of evidence has not yet emerged. Researchers suggest several factors that may play a role in the composition and development of the microbiota, including the method of delivery and the infant's diet (human milk or formula). It has been hypothesized that the infant's gestational age at birth also may play a role in the microbial inhabitants of its gut. Studies have shown that there are significant differences between the microbiomes of hospitalized, preterm infants and healthy, full-term babies (Palmer et al. 2007, 1557).

Palmer et al. studied the development of the gastrointestinal microbiome over the course of the first year of life for 14 healthy, full-term infants, including one set of fraternal twins. They also studied the infants' mothers by profiling their stool, vaginal, and milk samples and some of the infants' fathers and siblings by profiling their stool samples (Ibid.). Palmer et al. and his colleagues used microarray and sequencing to profile the 16S rRNA sequences from the participants' samples (Curtin 2009). They found that the microbiome develops quickly, noting that within a week of life, infants are colonized by a surprising number of taxa (Palmer et al. 2007). After one year of life, the infant's microbiome is "adult-like" and unique, when compared to the adults studied and the other children their own age (Ibid.). The microbiomes of the fraternal twins were not significantly distinct from one another during the first year (Ibid.). Palmer et al. and colleagues use the similarity of the fraternal twins' microbiome to suggest the strong influence of environment.

Palmer et al. argues that infants are first colonized by opportunistic microbes that are found in the environments to which they are exposed (Palmer et al. 2007, 1567). Unsurprisingly, a prominent source of microbiota for each baby is its mother (Ibid.). A major source of environmental microbes comes from the mother's vaginal, fecal, and skin microbiome (Ibid.). Dominguez-Bellow et al. have found that the types of bacteria that colonize newborn infants vary largely with mode of delivery (C-section and vaginal birth) (Dominguez-Bello et al. 2010). The first microbial exposures and colonizations that vaginally delivered infants have resemble their mother's vaginal microbiome, while infants delivered by C-section have early exposure and colonization that resembles their mother's skin microbiome. An obvious and yet to be answered question is whether differences in early colonization can have a long-acting influence on the subsequent development of the infant's "microbial succession patterns" in various body sites (Dominguez-Bellow et al. 2010, 11972). Early colonization may also have an impact on risk of infection by pathogenic microorganisms, which is generally higher in infants delivered through C-section (Ibid.). At the same time, the work by Palmer et al. suggests that the infant's microbiome is affected by whether the microbes find the infant's body a hospitable environment or not, and how some specific microbes fare in competition with other microbes. Over the first year of the infant's life, its gastrointestinal tract changes anatomically as does the infant's diet, as it changes from breast milk or formula to a more adult diet. The kinds of bacteria that endure are those that are most well adapted to this changed environment (Palmer et al. 2007, 1567).

The discussion of the infant's acquisition of microbiota has led many researchers to investigate whether or not there is a core microbiome, which is "the set of [microbial] genes present in a given habitat in all or the vast majority of humans" (Turnbaugh et al. 2007, 806). The core microbiome is contrasted with the "variable microbiome." A wide range of factors likely contribute to the diversity of the variable microbiome among individuals, including age, gender, "host genotype, host physiological status (including the properties of the innate and adaptive immune systems), host pathobiology (disease status), host lifestyle (including diet), host environment (at home and/or work), and the presence of transient populations of microorganisms that cannot persistently colonize a habitat" (Ibid.). Whether or not there is a core microbiome remains uncertain.

Before settling the question of whether there is a core microbiome, researchers must determine what counts as a core microbiome. First, should the core be defined at "the level" of genus, species, strain, genes, or "function"? (Friedrich 2008, 778). Second, what percentage of the population must share the same group of microorganisms for it to count as a core (Curtin 2009)? Third, is there a core microbiome for a limited geographic area, such as North America, or is there a core for the "global, continental, state, local, or some other region" (Ibid.)? Fourth, should the core be characterized in terms of similarity in abundance of a particular microbe, or is the mere presence of the microbial species among individuals sufficient?

Research thus far focusing on the microbial populations of various bodily locations and organ systems suggests that if there is a core microbiome that all humans share, it will have to be measured relative to a particular sample site (Hamady and Knight 2009, 1149). Qin et al. analyzed genetic material found in the fecal samples of 124 individuals and found that each individual "harbors at least 160 such species" of bacteria (Qin et al. 2010, p. 64). Considerable commonality was discovered; 75 bacterial species were common to over 50% of the individuals and 57 species common to over 90% of the individuals (Qin et al. 2010). Relman and Nelson, of the J. Craig Venter Institute, have investigated the core microbiome and the level of diversity of healthy adult microbiomes. In 2005, with the help of three healthy adult participants, Relman and Nelson examined samples of mucosa and fecal matter using 16s rDNA analysis. They discovered 244 new bacterial phenotypes; the majority of the organisms present in the samples were either of the phyla *Firmicutes* or *Bacteroides* (Curtin 2009). A second study in 2006 used shotgun sequencing to examine the microbial diversity of the fecal matter of two healthy adults (Ibid.).

The results of these and other studies, including the work of Turnbaugh and Hamady et al. on obese and lean twin pairs and their mothers, reveal that to a great extent family members share their gut microbiome (Turnbaugh, Hamady et al. 2008). At the level of microbial genes and their functions, researchers found that families shared many microbial genes that constituted "an extensive, identifiable 'core microbiome'" (Turnbaugh, Hamady et al. 2008, p. 2). At the level of the "bacterial lineage," however, researchers found significant diversity between family members, including twins. In the same study, researchers discovered that obesity was associated with divergence from the family's core microbiome, in terms of the phylum of bacteria present, bacterial diversity, and changes in both "bacterial genes and metabolic pathways" (Turnbaugh, Hamady et al. 2008, p. 2).

A recent study has shown that the skin microbiome on the hands is extremely variable across individuals, even suggesting that each individual has a unique microbial fingerprint that could be used in forensic identification (Fierer et al. 2010). By taking swabs of fingertips and palms, as well as surfaces with which individuals had regular contact, such as a computer keyboard or mouse, researchers retrieved microbial samples. They were able to use the "structure of the bacterial communities" that they found on the surfaces to correctly identify the individual who had contact with it. The core skin microbiome has been further studied by Grice et al. Taking samples from 10 healthy participants across 20 skin locations and over a period of four to six months, Grice et al. assessed microbial diversity along four dimensions: intrapersonal, interpersonal, temporal, and topographic (Grice et al. 2009). They found that variation between people was relative to the skin site that was being measured; they identified the spaces in between fingers and toes, the armpit, and the bellybutton as the most diverse between individuals and the back, outside of the nose, and nostrils as the most similar between individuals (Grice et al. 2009). The most dominant constituents of the skin microbiome are members of the group *Actinobacteria*, *Firmicutes*, and *Proteobacteria* (Grice et al. 2009, p. 2).

The level of global diversity of the human salivary microbiome has also been investigated (Nasidze et al. 2009). Because the mouth serves as a central entry point for microorganisms to access the rest of the human body, and because saliva swabs are an easily accessible and noninvasive way to sample large numbers of individual microbiomes, a number of studies have investigated saliva (Nasidze et al. 2009, 636). Researchers sampled 120 individuals from 12 sampling locations around the world, including North America, South America, Western Europe, Eastern Europe, Africa, and Eastern Asia. Over 600 species of bacteria have been discovered to inhabit the mouth. The most commonly detected bacterial genus worldwide, which alone accounted for 22.7% of the sequences, was *Streptococcus* (Nasidze et al. 2009, 637). Overall, a large level of diversity was found around the world. Based on the geographical origins of the samples, considerable differences were found in the population of particular microbes (Nasidze et al. 2009, 639). Researchers found that about 24% of interpersonal variation could be explained by distance from the equator. For the most part, there is very little overall "geographic patterning" of the salivary microbiome, although the salivary microbiome of participants from the Congo was particularly distinct" (Nasidze et al. 2009, 639).

Physicians and researchers have long recognized that the vaginal microbiome plays an important role in women's gynecological health. The composition of the vaginal microbiome is influenced by many factors, including, "age, hormonal fluctuations, sexual activity, [and] sanitary habits" (Badger et al. 2011, 8). Although Lactobacillus is often the most prevalent bacteria in healthy vaginal microbiome, recent studies suggest that this is not always the case and the composition of healthy vaginal microbiomes may be related to differences in ethnic groups (Hyman et al. 2005; Zhou et al. 2007).

Interactions Between the Microbiome and the Host Genome

The completion of the Human Genome Project between 2001 and 2003 was the beginning of a golden age in genetic research. In the following years, researchers discovered that the number of genes in the human genome is much lower than initially estimated and that there is an enormous amount of individual differences in DNA sequence and in copy number (deletions and duplications). The number of different proteins is significantly greater than the number of genes due to different splicing of coding areas (exons) within the genes, resulting in novel products. Researchers also discovered new kinds of small RNAs. Their functions included the regulation of gene activity and even the complete inactivation of genes.

A major mechanism for altering gene activity was shown to be methylation of DNA and of histones. These changes can be inherited over several generations without a change in the sequence of the DNA, originally thought to be the only mechanism (mutation) for inherited changes. Researchers have also begun to understand that alterations in the functions of genes result from mechanisms other than changes in the DNA sequence. This discovery spawned the field of epigenetics. Epigenetic phenomena involve "a change in phenotype that is heritable but does not involve DNA mutation" (Gottschling 2007, 2).

The human microbiome has been called our "other genome" (Zhao 2010). Some suggest that it is useful to think of our genome combined with the genome of our microbiome as a greater entity, the metagenome. However, the notion of a metagenome may be misleading. Variation in a microbiome genome is due not only to changes in its DNA (and in the case of many viruses, its RNA) but also to changes in the specific kinds of organisms that inhabit the human body. The development of accurate and reliable new techniques and tools for analyzing the genome of the microbiome will allow researchers to document genetic and species variation in the members of the microbiome across hosts (Zhao and Shen 2010). Discovery of these variations will help to clarify the role of the microbiome in human health and disease.

Scientists working in the field of immunology have also begun to better understand the ways in which human and animal hosts can resist pathological or harmful microorganisms as well as the mechanisms involved in increasing susceptibility to the diseases caused by the presence of these microscopic invaders. The mechanisms of host defense allow our bodies to defend against foreign substances and cells. These mechanisms are responsible for how our bodies react to invaders and how we differentiate our own cells from foreign cells. The two major components of the host defense mechanism are general innate immunity, which is a type of immunity that all members of a species share, and specific immunity, an individual's unique set of antibodies from previous exposures. Innate immunity is the function of a subset of immune cells (mostly lymphocytes) that recognize the presence of foreign organisms, such as certain bacteria, and aid in their destruction, providing an early defense against infection. Specific immunity is the far more complex form of immunity. In specific immunity lymphocytes and other cells cooperate to recognize and kill invading organisms by producing specific antibodies or by secreting various cell products that attack foreign cells. Sometimes, to the detriment of the host, these defensive cells and molecules attack the host, causing autoimmune diseases and inflammatory responses, including allergic reactions.

All of the molecular components of the host defense response are composed of proteins that are produced by genes. When the genes that direct protein production contain mutations, the protein activity is reduced or abolished, usually making the host more vulnerable to pathogenic organisms. It is also possible, however, that mutations in the genes that produce immunity-related proteins can leave a host more resistant to invasion. Mutations in the genes associated with the various components of the immune system cause a group of conditions called primary immunodeficiency diseases.

Over 100 of such diseases have now been recognized, each caused by mutations in genes responsible for the production of elements of the immune response, such as cytokines, immunoglobulins, and receptors responsible for the recognition of specific organisms (Notarangelo 2010). Mutations in these genes may cause the host to develop an immunodeficiency disease.

It should be noted that both the microbiome and invasive organisms consist of not only bacteria but also viruses, fungi, protozoa, and even multicellular organisms such as parasites. A number of interactions between microbes and human genes and their combined relationship to mechanisms of susceptibility and resistance to various bacteria and viruses have recently been uncovered. Much of this work has been done by Alcaïs et al., as well as an increasing number of other investigators (see Alcaïs, Abel, and Casanova 2009). A few examples will illustrate the importance of this type of research.

Tuberculosis is the most common and serious form of infection with mycobacteria. Infection with mycobacteria has devastating effects on the health of some individuals and is essentially asymptomatic in others. This difference is due to genetic variation in the infected individuals (Al-Muhsen and Casanova 2008).

Similarly, infants who cannot produce or have a deficiency in the molecule called UNC-93B, due to a genetic mutation, suffer a marked increase in susceptibility to herpes simplex virus encephalopathy (Casrouge et al. 2006). Comparable relationships have been found in animal models of human diseases, many of which will undoubtedly lead to the identification of identical or similar mutations in humans.

For example, a study done on mice that have a genetic mutation that makes them more susceptible to autoimmune arthritis showed that when their intestines were also colonized with a specific segmented filamentous bacteria, they would not develop the symptoms of arthritis despite having the genetic mutation (Wu et al. 2010). Similar results show that mice that are genetically susceptible to inflammatory bowel disease (IBD) did not develop the disease when their guts were colonized exclusively with the bacterium *Bacteroides fragilis* (Round and Mazmanian 2010). In another mouse study, findings suggests that mice lacking a gene important in natural immunity (Toll-like receptor 5) develop metabolic syndrome (obesity and type 2 diabetes) when their intestinal microbiome is altered (Vijay-Kumar et al. 2010).

Other reports show the complexity of genome–microbiome interactions. In studies using a mouse model, mice were given a mutated gene associated with Crohn's disease susceptibility. This mutation was originally detected in humans by genome-wide association studies. Researchers found that the disease did not develop unless two conditions were met: their intestines had to be colonized by specific bacteria and they had to be infected with a subtype of norovirus (a cause of diarrhea in humans) (Cadwell et al. 2010). These studies demonstrate the high level of complexity in interactions between host genes, intestinal bacteria, and viruses.

Encouraged by such findings, Khoruts et al. have treated a group of patients infected with the bacterium *Clostridium difficile* with transplantation of feces from normal donors (Khoruts et al. 2010; McKenna 2011). *C. difficile,* which is not normally part of the human microbiome, causes a debilitating and often lethal disease. Infection with *C. difficile* often develops after prolonged antibiotic treatment. The fecal transplantation procedures were performed either orally or through a colonic infusion. The treatment cured the infection by colonizing the patient with several different species of bacteria found in the donated fecal transplant.

Changes in the genes of the microbes that inhabit our intestines can also alter the microbial–human relationship. For example, a mutation in *E. coli* bacteria, which commonly colonize our intestines, can cause intestinal inflammation and may enhance susceptibility to colon cancer (Cuevas-Ramos et al. 2010). An example of increase in resistance to human disease is the finding that the human gene responsible for susceptibility to celiac disease simultaneously produces increased resistance to bacterial infection. This may explain the high incidence of celiac disease, which causes diarrhea, malabsorption, and, ultimately, destruction of the intestinal lining. Similarly, a recent study by Olaszak and colleagues has provided support for the "hygiene hypothesis" that posits negative effects from an overly aggressive battle against germs (Olszak et al. 2010). These findings suggest that the diversity of microbes that we encounter early in life provide protection against inflammatory and autoimmune conditions such as allergies, asthma, IBD, and multiple sclerosis (Leslie 2012). Exploration of the complex relationship between the human genome, microbiome, and metagenome is expanding rapidly and is expected to lead to enhanced understanding of human health, disease, and potential treatments.

Manipulating the Microbiome for Medical Purposes

There is increasing evidence that changes in the human microbiome are associated with human health and disease. Again, research initiatives like the HMP are seeking to characterize the microorganisms that colonize various body sites, most notably the gut, the genitals, the skin, and the mouth. It may become possible to cure or prevent diseases like diabetes or sexually transmitted infections by intentionally manipulating or mimicking the role of the healthy human microbiome.

Indeed, even though scientific understanding of the composition, role, and function of the human microbiome is still limited, numerous studies of this kind are already under way. These efforts range from studies examining the use of probiotics and dietary interventions to treat common digestive disorders to plans for colonizing animal models with genetically modified organisms.

In this section we will review some of the current and planned efforts by focusing on two areas of intense research: manipulating the intestinal microbiome to prevent or cure obesity and other digestive disorders and altering the vaginal microbiome to prevent or treat sexually transmitted infections.

Studies suggest that the microbiome plays a role in gastrointestinal disorders and associated diseases such as IBD, dietary allergy, and obesity (c.f. Backhed and Gordon et al. 2007; Blaut and Clavel 2007; Tsai and Coyle 2009).

It is not surprising, then, that a number of clinical trials are already under way to explore ways of manipulating the intestinal microbiome as a treatment strategy. The most common of these approaches involves the use of either prebiotics or probiotics. Prebiotics are nondigestible food ingredients, such as soluble fiber, that stimulate the growth or activity of microorganisms in the gut. Probiotics are themselves microorganisms; they are contained in many commonly fermented foods including, most notably, yogurt and other cultured dairy products. Both prebiotics and probiotics are believed to work by promoting the growth of healthy bacteria and yeasts while inhibiting pathogens or other harmful microorganisms from colonizing the gastrointestinal tract. The regular use of prebiotic supplements and probiotic foods is heavily marketed in the popular press as a way of maintaining or improving health. Prebiotics and probiotic foods and supplements to promote weight loss, treat colds and allergies, prevent cancer, and even delay the aging process are already available commercially, although few such foods and supplements have been shown conclusively to achieve such goals.

The most advanced research in this area involves manipulation of the intestinal microbiome to treat or prevent IBD. In an article recently published in *Current Gastroenterology Reports*, for example, Gulati and Dubinsky reviewed published trials studying probiotics for the treatment of gastrointestinal disorders like IBD, with particular emphasis on their role in pediatric patients (Gulati and Dubinsky 2009).

A number of groups also are exploring the use of prebiotics and probiotics to prevent or treat dietary allergies. In a recent meta-analysis, Osborn and Sinn assessed seven trials that explored the effect of prebiotics and probiotics on allergic disease outcome in infants (Osborn and Sinn 2008). In five studies, they found that the use of a *Lactobacillus*-containing probiotic supplement was associated with a significant reduction in infant eczema (painful and itchy skin rash triggered by exposure to allergens). Similarly, a study of prebiotic supplementation among exclusively formula-fed infants reported a similar reduction in eczema, but the utility of this study was limited by the small sample size and large number of participants lost to follow-up.

Finally, several studies have described the role of intestinal microorganisms in the regulation of energy homeostasis, both in animal models and in humans. Bäckhed and his colleagues, for example, found that mice raised in a germ-free environment (and thus free of the conventional intestinal flora that colonize normal mice) have lower overall body fat content (Backhed and Gordon, et al. 2004; Bäckhed,, Fredrik and Gordon et al., 2007). When later colonized with gut microorganisms extracted from normal mice, these same germ-free mice increased their fat mass by 60% (Turnbaugh et al. 2006). Other studies have shown that the intestinal ecology of obese humans also differs from their leaner counterparts. Obese individuals, for example, have greater numbers of organisms from the *Firmicutes* family and fewer from the *Bacteroides* family of bacteria. When 12 obese study participants were placed on a restricted diet, their intestinal microbiota changed to be more similar to what is seen in normal-weight individuals. Several groups are currently exploring the use of prebiotics to induce changes in the intestinal microbiome to treat obesity and other metabolic disorders (Turnbaugh, Hamady et al. 2008).

The vagina in healthy women of childbearing age is colonized by a variety of mutualistic microorganisms, particularly *Lactobacillus*. Lactobacilli that are found in the vagina produce a number of compounds that inhibit pathogenic microorganisms. These compounds include lactic acid, hydrogen peroxide, lactacin, and acidolin. These compounds also maintain the acidity of the vagina. It is believed that these antimicrobial compounds, coupled with the acidity of the vagina, help to prevent the acquisition of common sexually transmitted diseases (STDs) like HIV, gonorrhea, and bacterial vaginosis.[7]

Reproductive health researchers are currently trying to develop treatments designed to maintain or restore the normal, healthy vaginal ecology and to treat or prevent STDs. For example, researchers have developed and are testing a suppository containing *Lactobacillus crispatus*. Early phase I/II studies have found that regular insertion of the suppository was safe. It also allowed hydrogen peroxide–producing Lactobacilli to become part of the vaginal flora. Hydrogen peroxide is a naturally occurring antimicrobial compound, and, as mentioned previously, it is believed that hydrogen peroxide–producing bacteria prevent more pathogenic organisms from colonizing the vagina and causing disease. Further studies are planned to test the suppository as a treatment for or prophylactic against a variety of STDs. *L. crispatus* is the dominant organism in about 20% of women and is present in nearly all women. Researchers have developed a similar *L. crispatus*–containing suppository called Lactin-V. It is currently in phase II trials as a treatment for recurrent urinary tract infections and recurrent bacterial vaginosis.

There are also plans to test the use of genetically modified *Lactobacillus*, which produce specific antimicrobials like cyanovirin-N to treat and prevent genital infections. Researchers have inserted the gene for cyanovirin-N, a potent in vitro inhibitor of HIV, into *Lactobacillus* and, based on pharmacokinetic data from primate studies, expect to begin clinical trials soon (Xiaowen et al. 2006).

Conclusion

Technological developments and scientific advances in understanding the human microbiome and its complex role in maintaining health and causing

---

[7] A subgroup of women have vaginal microbiota dominated by a mixture of organisms other than lactobacilli.

disease are occurring rapidly. Studies that were under way when this chapter was written may have been superseded by the time of its publication. Despite our efforts to make this chapter current, readers should be alert to how quickly the landscape in this domain is changing.

This chapter was also intended to provide a very broad overview of the HMP and current research on the human microbiome. It offers a sampling of the types of research questions that scientists working in this field are asking, rather than a comprehensive review. Our aim was to provide the reader with a sense of the range and scope of this research and an overview of how this work is related to other branches of biology and medicine, including genetics, immunology, gastroenterology, metabiology, and so on. Second, the chapter was intended to convey the potentially significant impact that scientific findings related to the microbiome may have on our health and the practice of medicine. Third, the chapter aims to encourage readers to reconsider the picture of a combative relationship between microbes and their human hosts. Sophisticated physicians and researchers have long been aware of the beneficial role of the normal microflora in human health. Today's microbiome research may improve the popular press and public understanding and help them not only to appreciate the role of microbes in allowing our bodies to function, but also to understand just how closely intertwined our biological lives are with our environment, including vast numbers of microorganisms.

References

Abraham, E. P., and E. Chain. 1940. Letters to Editor, an enzyme from bacteria able to destroy penicillin. *Nature* 146: 837.

Ackerman, Jennifer. 2012. The ultimate social network. *Scientific American* 306(6): 37–43.

Alami, Younes, Wafa Achouak, Christine Marol, and Thierry Heulin. 2000. Rhizosphere soil aggregation and plant growth promotion of sunflowers by an exopolysaccharide-producing Rhizobium sp. strain isolated from sunflower roots, PGPR: biocontrol and biofertilization. *Applied and Environmental Microbiology* 66(8): 3393–3398.

Alcaïs, A., L. Abel, and J. L. Casanova. 2009. Human genetics of infectious diseases: between proof of principle and paradigm. *Journal of Clinical Investigation* 119(9): 2506–2514.

Al-Muhsen, Saleh, and Jean-Laurent Casanova. 2008. The genetic heterogeneity of mendelian susceptibility to mycobacterial diseases. *Journal of Allergy and Clinical Immunology* 122(6): 1043–1051.

Amann, R., W. Ludwig, and K. H. Schleifer. 1995. Phylogenetic identification and in situ detection of individual microbial cells without cultivation. *Microbiology Review* 59(1): 143–169.

Arumugam, Manimozhiyan, et al. 2011. "Enterotypes of the human gut microbiome." *Nature* 473: 174–180.

Bäckhed, Fredrik and Gordon, Jeffrey I., et al. 2004. "The gut microbiota as an environmental factor that regulates fat storage." *PNAS* 101(44): 15718.

Bäckhed, Fredrik and Gordon, Jeffrey I., et al. 2007. Mechanisms underlying the resistance to diet-induced obesity in germ-free mice. *Proceedings of the National Academy of Sciences of the United States of America* 104(3): 979–984.

Badger, Jonathan H., Pauline Ng, Craig Venter, et al. 2011. The human genome, microbiomes, and disease. In Karen E. Nelson (ed.), *Metagenomics of the Human Body* (p. 8). New York: Springer.

Baldwin, P. 2005. *Contagion and the State in Europe,* 1830—1930. Cambridge: Cambridge University Press, 2005.

Barket, Theo, and Michael Drake, eds. 1982. *Population and Society in Britain* 1850-1980. New York: New York University Press.

Berkner, L. V., and L. C. Marshall. 1965. On the origin and rise of oxygen concentration in the earth's atmosphere. *Journal of the Atmospheric Sciences* 22(3): 225–261.

Blaser, M. J., and J. C. Atherton. 2004. Helicobacter pylori persistence: biology and disease. *Journal of Clinical Investigation* 113(3): 321–333.

Blaser, M. J., and S. Falkow. 2009. What are the consequences of the disappearing human microbiota? *Nature Reviews Microbiology* 7: 887–894.

Blaser, Martin J. 1997. Perspectives series: host/pathogen interactions: ecology of Helicobacter pylori in the human stomach. *Journal of Clinical Investigation* 100(4): 759–762.

Blaser, Martin J. 2006. Pathogenicity and symbiosis: human gastric colonization by Helicobacter pylori as a model of amphibiosis. In National Research Council. *Ending the War Metaphor: The Changing Agenda for Unraveling the Host-Microbe Relationship.* Washington, DC: National Academies Press: 115–140.

Blaser, Martin J., Yu Chen, and Joan Reibman. 2008. Does Helicobacter pylori protect against asthma and allergy? *Gut* 57(5): 561–567.

Blaut, Michael, and Thomas Clavel. 2007. Metabolic diversity of the intestinal microbiota: implications for health and disease. *Journal of Nutrition* 137(3): 751S–755S.

Breitbart, M., et al. 2002. Genomic analysis of uncultured marine viral communities. *Proceedings of the National Academy of Sciences of the United States of America* 99(22): 14250–14255.

Brooks, G. F., K. C. Carroll, J. S. Butel, and S. A. Morse. Normal microbial flora of the human body. In *Jawetz, Melnick, & Adelberg's Medical Microbiology* (24th ed.). http://www.accessmedicine.com/content.aspx?aID=2756528.

Butcher, S. S., R. J. Charlson, G. H. Orians, and G. V. Wolfe (eds.). 1992. *Global Biogeochemical Cycles.* London: Academic Press.

Cadwell, Ken, et al. 2010. Virus-plus-susceptibility gene interaction determines Crohn's disease gene Atg16L1 phenotypes in intestine. *Cell* 141(7): 1135–1145.

Casrouge, Amanda, et al. 2006. Herpes simplex virus encephalitis in human UNC-93B deficiency. *Science* 314: 308–312.

Chivian, D., Brodie, E. L., et al. 2008. Environmental genomics reveals a single-species ecosystem deep within Earth. *Science* 322(5899): 275–278.

Clark, Ronald W. 1985. *The Life of Ernst Chain Penicillin and Beyond.* New York: St. Martin's Press.

Cook, G. C., A. I. Zumla, and P. Manson (eds.). 2009. *Manson's Tropical Diseases.* Edinburgh: Saunders-Elsevier.

Cuevas-Ramos, Gabriel, Claude R. Petit, Ingrid Marc, Michèle Boury, Eric Oswald, and Jean-Philippe Nougayrède. 2010. Escherichia coli induces DNA damage in vivo and triggers genomic instability in mammalian cells. *Proceedings of the National Academy of Sciences of the United States of America* 107(25): 11537–11542.

Curtin, C. 2009. Humans as host. *Genome Technology Newsletter* 95: 38–43.

De Kruif, P. 1926. *Microbe Hunters*. New York: Harcourt Book.

DeWeerdt, Sarah, A. 2002. The world's toughest bacterium. *Genome News Network*. http://www.genomenewsnetwork.org/articles/07_02/deinococcus.shtml. Accessed November 3, 2009.

Dinsdale, E. A., et al. 2008. Functional metagenomic profiling of nine biomes. *Nature* 452(7187): 629–632.

Dobell, C. 1932. *Anton van Leeuwenhoek His Little Animals*. New York: Dover Publications.

Dominguez-Bello, Maria G., Elizabeth K. Costello, Monica Contreras, Magda Magris, Glida Hidalgo, Noah Fierer, and Rob Knight. 2010. Delivery mode shapes the acquisition and structure of the initial microbiota across multiple body habitats in newborns. *Proceedings of the National Academy of Sciences of the United States of America* 107(26): 11971–11975.

Dubos, Rene. 1960. *Loius Pasteur: Free Lance of Science*. Boston: Little, Brown and Company.

Ellis, Harold. 2001. *A History of Surgery*. Grafos, Spain: Greenwich Medical Media Limited.

Emerson, David, and Criag L. Moyer. 2002. Neutrophilic Fe-oxidizing bacteria are abundant at the Loihi Seamount hydrothermal vents and play a major role in Fe oxide deposition. *Applied and Environmental Microbiology* 68(6): 3085–3093.

Environmental Protection Agency. n.d. *The citizens guide to bioremediation*. http://www.clu-in.org/download/citizens/bioremediation.pdf. Accessed November 3, 2009.

Falk, P. G., L. V. Hooper, T. Midtvedt, and J. L. Gordon. 1998. Creating and maintaining the gastrointestinal ecosystem: what we know and need to know from gnotobiology. *Microbiology and Molecular Biology Reviews* 62(4): 1157–1170.

Falkow, Stanley. 2006. Is persistent bacterial infection good for your health? *Cell* 124(4): 699–702.

Feldman, Robert, A. 2004. Coevolution of Symbionts and Pathogens with their Hosts. In Fraser, C. M., T. D. Read, and K. E. Nelson (eds.), *Microbial Genomes* (pp. 195–213). Totowa, NJ: Humana Press.

Fierer, N., M. Hamady, C. L. Lauber, and R. Knight. 2008. The influence of sex, handedness, and washing on the diversity of hand surface bacteria. *Proceedings of the National Academy of Sciences of the United States of America* 105(46): 17994–17999.

Fierer, Noah, Christian L. Lauber, Nick Zhou, Daniel McDonald, Elizabeth K. Costello, and Rob Knight. 2010. Forensic identification using skin bacterial communities. *Proceedings of the National Academy of Sciences of the United States of America* 107(14): 6477–6481.

Fischer, S. G., and L. S. Lerman. 1979. Length-independent separation of DNA restriction fragments in two-dimensional gel electrophoresis. *Cell* 1: 191–200.

Foxman, B., D. Goldberg, C. Murdock, C. Xi, and J. R. Gilsdorf. 2008. Conceptualizing human microbiota: from multicelled organ to ecological community. *Interdisciplinary Perspectives on Infectious Diseases* 2008: 613979.

Frank, D. N., and N. R. Pace. 2008. Gastrointestinal microbiology enters the metagenomics era. *Current Opinion in Gastroenterology* 24: 4–10.

Friedrich, M. J. 2008. Microbiome project seeks to understand human body's microscopic residents. *Journal of the American Medical Association* 300(7): 777–778.

Fuentes-Ramirez, L. E. and Caballero-Mellado S. 2005. Bacterial biofertilizers. In Z. A. Siddiqui (ed.), *PGPR: Biocontrol and Biofertilization* (pp. 143–172). Dordrecht, The Netherlands: Springer.

Goodacre, R. 2007. Supplement: International Research Conference on Food, Nutrition, and Cancer Metabolomics of a Superorganism. *American Society for Nutrition Journal of Nutrition* 137: 259S–266S.

Gottschling, Daniel E. 2007. Epigenetics from phenomenon to field. In Allis, C. D., Jenuwein, T., Reinberg, D., and Caparros, M. L. (eds.), *Epigenetics* (pp. 1–15). Cold Spring Harbor, NY: Cold Spring Harbor Laboratory Press.

Grice, Elizabeth A., and Julia A. Segre. 2011. The skin microbiome. *Nature Reviews Microbiology* 9: 244–253. doi:10.1038/nrmicro2537.

Grice, Elizabeth, et al. 2009. Topographical and temporal diversity of the human skin microbiome. *Science* 324(5931): 1190–1192.

Gulati, A. S., and M. C. Dubinsky. 2009. Probiotics in pediatric inflammatory bowel diseases. *Current Gastroenterology Reports* 11(3): 238–247.

Hamady, M., and R. Knight. 2009. Microbial community profiling for human microbiome projects: tools, techniques, and challenges. *Genome Research* 19: 1141–1152.

Handelsman, J., M. R. Rondon, S. F. Brady, J. Clardy, and R. M. Goodman. 1998. Molecular biological access to the chemistry of unknown soil microbes: a new frontier for natural products. *Chemistry and Biology* 5(10): R245–R249.

Heritage, J., E. G. V. Evans, and R. A. Killington. 1999. *The Microbiology of Soil and Nutrient Cycling: Microbiology in Action*. Cambridge, United Kingdom: Cambridge University Press.

Honey, K. 2008. Good bugs, bad bugs: learning what we can from the microorganisms that colonize our bodies. *Journal of Clinical Investigation* 118(12): 3817.

Hopkins, D. R. 2002. *The Greatest Killer: Smallpox in history*. United States of America: University of Chicago Press. .

Hyman, R.W., et al. 2005. Microbes on the human vaginal epithelium. *Proceedings of the National Academy of Sciences of the United States of America* 102(22): 7952.

Kasting, J. F. 2005. Methane and climate during the Precambrian era. *Precambrian Research* 137(3–4): 119–129.

Khoruts, A., J. Dicksved, J. K. Jansson, et al. 2010. Changes in the composition of the human fecal microbiome after bacteriotherapy for recurrent Clostridium difficile-associated diarrhea. *Journal of Clinical Gastroenterology* 44: 354–360.

Kingston, W. 2004. Streptomycin, Schatz v. Waksman, and the balance of credit for discovery. *Journal of the History of Medicine and Allied Sciences* 57(3): 441–462.

Kinross, J. M., A. C. von Roon, E. Holmes, A. Darzi, and J. K. Nicholson. 2008. The human gut microbiome: implications for future health care. *Current Gastroenterology Reports* 10: 396–403.

Krimsky, Sheldon and Wrubel, Roger. 1996. *Agricultural Biotechnology and the Environment: Science, Policy and Social Issues*. United States of America: University of Illinois Press.

Lane, D. J., et al. 1985. Rapid determination of 16S ribosomal RNA sequences for phylogenetic analyses. *Proceedings of the National Academy of Sciences of the United States of America* 82(20): 6955–6959.

Lederberg, Joshua. Summer 2003. Of men and microbes. *New Perspectives Quarterly* 2120(3): 53–55.

Leslie, Mitch. 2012. Gut microbes keep rare immune cells in line. *Science* 23(335): 1428.

Levinson, W. n.d. Normal flora. In *Review of Medical Microbiology and Immunology* (10th ed.). United States of America: The McGraw Hill Companies, Inc. http://www.accessmedicine.com/content.aspx?aID=3334958. Accessed November 5, 2009.

Lister, Joseph. 1867. On the antiseptic principle in the practice of surgery. *British Medical Journal* 2(351): 246–248.

Maczulak, Anne. 2011. *Allies and Enemies: How the World Depends on Bacteria*. Upper Saddle River, New Jersey: Pearson Education, Inc.

Maxam, A. M., and W. Gilbert. 1977. A new method for sequencing DNA. *Proceedings of the National Academy of Sciences of the United States of America* 74(2): 560–564.

McKenna, Maryn. 2011. Swapping germs: a potentially beneficial but unusual treatment for serious intestinal ailments may fall victim to regulatory difficulties. *Scientific American* 305(6): 34–36.

Minor, P. D. 2007. Viruses. *Encyclopedia of Life Sciences*. doi:10.1002/9780470015902.a0000441.pub2 (Article Online Posting Date: December 21, 2007).

Mullard, A., and A. Mandavilli. 2008. Microbiology: the inside story. *Nature* 453(7195): 578–580.

Mullard, A., and A. Mandavilli. *The New Science of Metagenomics: Revealing the Secrets of Our Microbial Planet* (p. 37).

Nasidze, I., J. Li, D. Quinque, K. Tang, and M. Stoneking. 2009. Global diversity in the human salivary microbiome. *Genome Research* 4: 636–643. Epub 2009 Feb 27.

National Institutes of Health. n.d.-a. *Roadmap for Medical Research, Human Microbiome Project*. Brainstorming. National Institutes of Health, U.S. Department of Health and Human Services. http://nihroadmap.nih.gov/hmp/. Accessed November 11, 2009.

National Institutes of Health. n.d.-b. *Roadmap for Medical Research, Human Microbiome Project*. Overview. National Institute of Health, U.S. Department of Health and Human Services. Available: http://nihroadmap.nih.gov/hmp/. Accessed November 2009.

National Research Council. 2007. *The New Science of Metagenomics: Revealing the Secrets of Our Microbial Planet*. Washington, DC: The National Academies Press. Available: http://www.nap.edu/openbook.php?record_id=11902&page=1

Niven, Charles F., Jr. n.d. Sterilization. In AccessScience@McGraw-Hill. http://www.accessscience.com. doi:10.1036/1097-8542.655600.

Notarangelo, Luigi D. 2010. Primary immunodeficiencies. *Journal of Allergy and Clinical Immunology* 125(2, Suppl. 2): S182–S194.

Olszak, Torsten, et al. 2012. Microbial exposure during early life has persistent effects on natural killer t cell function. *Science* 336(6080): 489–493.

Office of Science & Technology Policy and the National Science & Technology Council, Committee on Science, Subcommittee on Biotechnology. n.d. The Microbe Project. http://www.microbeproject.gov/. Accessed November 3, 2009.

Osborn, D., and J. Sinn. 2008. Chapter 9: What evidence supports dietary intervention to prevent infant allergy and food allergy development? In Robin K Ohls and Akhil Maheshwari (eds.), *Hematology, Immunology and Infectious Disease: Neonatology Questions and Controversies: Expert Consult – Online and Prin*t (2nd ed., pp. 11–128).

Ouwehand, Arthur C., and Elaine E. Vaughan (eds.). 2006. *Gastrointestinal Microbiology.* New York: Taylor & Francis Group.

Palmer, C., E. M. Bik, D. B. DiGiulio, D. A. Relman, and P. O. Brown. 2007. Development of the human infant intestinal microbiota. *PLoS Biology* 5(7): e177.

Pavlov, A. A., and J. F. Kasting. 2002. Mass-independent fractionation of sulfur isotopes in archean sediments: strong evidence for an anoxic archean atmosphere. *Astrobiology* 2(1): 27–41.

Pommerville, Jeffrey. 2011. *Alcamo's Fundamentals of Microbiology* (9th ed.). United States: Jones and Bartlett Publishers.

Qin, Junjie, et al. 2010. A human gut microbial gene catalogue established by metagenomic sequencing. *Nature* 464: 59–65.

Riley, M. A., and M. A. Chavan. 2007. *Bacteriocins: Ecology and Evolution*: Heidelberg, Germany: Springer.

Risser, P. Biome. In AccessScience@McGraw-Hill. http://www.accessscience.com. doi10.1036/1097-8542.083400. Accessed June 23, 2012.

Round, J.L. and Mazmanian, S. K. 2010. Inducible Foxp3+ regulatory T-cell development by a commensal bacterium of the intestinal microbiota. *Proceedings of the National Academy of Sciences of the United States of America* 107: 12204–12211.

Sanger, F., and A. R. Coulson. 1975. A rapid method for determining sequences in DNA by primed synthesis with DNA polymerase. *Journal of Molecular Biology* 94(3): 441–448.

Saunders, P. 1982. *Edward Jenner: The Cheltenham Years 1795-1823*. Hanover, NH: University Press of New England.

Schwartzman, D. W. 1999. *Life, Temperature, and the Earth.* New York: Columbia University Press.

Sleigh, Michael A. 2006. Protozoa. *Encyclopedia of Life Sciences.* doi:10.1038/npg.els.0004346 (Article Online Posting Date: January 27, 2006).

Smil, Vaclav. 2002. *The Earth's Biosphere.* United States of America: MIT Press.

Sogin, M. L., et al. 2006. Microbial diversity in the deep sea and the underexplored 'rare biosphere'. *Proceedings of the National Academy of Sciences of the United States of America* 103(32): 12115–12120.

Szewzyk, U., Szewzyk, R., and Stenström, T. A. 1994. Thermophilic, anaerobic bacteria isolated from a deep borehole in granite in Sweden. *Proceedings of the National Academy of Sciences of the United States of America* 91(5): 1810–1813.

Tannock, G. 1994. *Normal microflora: an introduction to microbes inhabiting the human body*. London, United Kingdom: Springer.

Thomas, P. (ed.). Harnessing invisible power. In *Lucent Library of Science and Technology: Bacteria and Viruses* (pp. 67–80). San Diego, CA: Lucent Books.

Thompson, Robert Edward Mervyn. 1963. Cross infection in hospital. *New Scientist* 3451.

Tringe, S. G., et al. 2005. Comparative metagenomics of microbial communities. *Science* 308(5721): 554–557.

Tsai, Franklin, and Walter J. Coyle. 2009. The microbiome and obesity: is obesity linked to our gut flora? *Current Gastroenterology Reports* 11(4): 307–313.

Turnbaugh, P. J., M. Hamady, T. Yatsunenko, B. L. Cantarel, A. Duncan, R. E. Ley, et al. 2008. A core gut microbiome in obese and lean twins. *Nature* 457: 480–484.

Turnbaugh, P. J., R. E. Ley, M. Hamady, C. M. Fraser-Liggett, R. Knight, and J. I. Gordon. 2007. The Human Microbiome Project. *Nature* 449: 804–810.

Turnbaugh, Peter J., Fredrik Bäckhed, Lucinda Fulton, and Jeffrey I. Gordon. 2008. Diet-induced obesity is linked to marked but reversible alterations in the mouse distal gut microbiome. *Cell Host and Microbe* 3(4): 213–223.

Turnbaugh, Peter J., et al. 2006. An obesity-associated gut microbiome with increased capacity for energy harvest. *Nature* 444: 1027–1031.

U.S. Food and Agriculture Administration and World Health Organization. Guidelines for the Evaluation of Probiotics in Food. London, Ontario, Canada, April 30 and May 1, 2002.

Van Epps, Heather. 2006. René Dubos: unearthing antibiotics. *Journal of Experimental Medicine* 203(2): 259.

Venter, J. C., et al. 2004. Environmental genome shotgun sequencing of the Sargasso Sea. *Science* 304(5667): 66–74.

Vijay-Kumar, M., et al. 2010. Metabolic syndrome and altered gut microbiota in mice lacking Toll-like receptor 5. *Science* 328: 228–231.

Whitman, B. W., D. C. Coleman, and W. J. Wiebe. 1998. Prokaryotes: the unseen majority. *Proceedings of the National Academy of Sciences of the United States of America* 95(12): 6578–6583.

Woese, C. R., and G. E. Fox. 1977. Phylogenetic structure of the prokaryotic domain: the primary kingdoms. *Proceedings of the National Academy of Sciences of the United States of America* 74: 5088–5090.

Wu, Hsin-Jung, et al. 2010. Gut-residing segmented filamentous bacteria drive autoimmune arthritis via T helper 17 cells. *Immunity* 32(6): 815–827.

Xiaowen, Liu, et al. 2006. Engineered vaginal lactobacillus strain for mucosal delivery of the human immunodeficiency virus inhibitor cyanovirin-N. *Antimicrobial Agents and Chemotherapy* 50(10): 3250–3259.

Xu, Jianping. 2010. Metagenomics and ecosystems biology: conceptual frameworks, tools and methods. In Diana Marco (ed.), *Metagenomics: Theory, Methods and Applications* (pp. 1–2). Norfolk, UK: Caister Academic Press.

Zhao, Liping and Shen, Jian. September 2010.Whole-body systems approaches for gut microbiota-targeted, preventive healthcare. *Journal of Biotechnology* 149(3): 183–190.

Zhao, Liping, 2010. Genomics: The tale of our other genome. *Nature* 465 (7300):879–880.

Zhou, X., et al. 2007. Differences in the composition of vaginal microbial communities found in healthy Caucasian and black women. *ISME Journal: Multidisciplinary Journal of Microbial Ecology* 1(2): 121–133.

2

Personal Identity

OUR MICROBES, OURSELVES

Nada Gligorov, Jody Azzouni, Douglas P. Lackey, Arnold Zweig

Introduction

People often speak of a dichotomy between scientific and commonsense views. Such dichotomies are conspicuous where the domains of scientific inquiry and common sense overlap. Fields such as social science, psychology, and even neuroscience seek to explain phenomena that are also within the purview of common sense. For example, conceptions of personhood, personal identity, autonomy, and other related concepts define who we are and are used to explain our ability to be moral. Conflicts between science and common sense arise when the two views provide diverging explanations of what seem to be the same concepts. For example, in everyday life we often rely on and invoke the concept of autonomy, while research in psychology sometimes points to our limited ability to make voluntary decisions and to reason free from error. In such conflicting cases, it may seem that only one view is correct and the other must be abandoned.

The apparent dichotomy, however, is false. It wrongly characterizes the relationship between science and common sense as a rivalry. Commonsense views are not independent from science; in fact, they are a patchwork of beliefs derived from diverse sources, including science. Moreover, beliefs about ourselves, such as those pertaining to the nature of personhood, personal identity, and autonomy, change over time, and have changed as a result of scientific inquiry. The true character of the relationship between science and common sense involves bidirectional influence and continuity.

Given that science has shaped, and continues to shape, conceptions of who we are, it is fitting to evaluate whether the Human Microbiome Project (HMP) will influence notions of personal identity. The HMP invites us to consider whether learning about the microbes that live on us and in us will influence our conception of who we are. In that light, we shall review various traditional criteria of personal identity and assess whether any of them would be affected by the conceptual shift that this new area of scientific discovery introduces. For the most part, we argue that numerical criteria for personal identity over time will not be significantly affected by discoveries related to the human microbiome.

In contrast to philosophical conceptions of personal identity, common-sense conceptions of personal identity and personhood are changeable and often influenced by personal experience or the acquisition of new knowledge. Hence, we shall argue that individual conceptions of the self could be affected by the popularization of the HMP. We could come to believe that features of our microbiome are features of ourselves. Thus, in this chapter we propose that the dissemination of scientific facts about the HMP be done in a way that is mindful of the influence that science has on common sense and how scientific language and advances can shape common sense and thereby have an impact on research participation and medicine.

Personal Identity over Time

To ascertain the relevance of the HMP to conceptions of personhood and personal identity, we shall start by considering some of the most frequently proposed criteria for personal identity. We begin by defining the problem of personal identity over time and then review the various criteria proposed to establish identity over time. We choose to discuss this topic because of its importance in the philosophical literature, and because it is not likely to be addressed by others working on the HMP. We review all the relevant conceptions of identity, ranging from philosophical to common sense, to explain which ones will be affected by research on the human microbiome.

The personal identity problem in philosophy centers on establishing a numerical criterion of identity over time, which means establishing that an object is one and the same object despite change that occurs over time. Consider the changes that occur during the lifespan of someone as they develop from childhood to old age. Looking through one of my albums, a friend inquired about a picture of a girl sitting at a piano. The picture was old and faded because it had been taken before World War II. I identified the person in the photograph as my grandmother. In thinking about my answer, I realized that it was quite misleading. The young girl in the picture was at most 12 years old and certainly was not anybody's grandmother. Was my answer incorrect? In what sense is the girl in the faded photograph *my* grandmother? Obviously, there is some relationship of continuity between the budding pianist in the photograph and my grandmother, and most people would probably regard my utterance as accurate. Yet there is still a sense in which the girl in the photograph is not the old woman who is now my grandmother. Her appearance has changed, as have her height, weight, and various other physical features. She has also endured large shifts in her beliefs and values and the basic features of her psychology. At the age of 12, she dreamt of being a musician; later on, she opted for the more practical profession of medicine. The personal identity problem is the problem of fixing the identity of my grandmother across the various stages of her life. Analogously, the problem arises for any person who endures changes in properties across time.

Broadly speaking, there are two main approaches to solving this problem within the philosophical literature; one approach uses a physical criterion to establish identity, while the other employs a psychological criterion. For both criteria, establishing personal identity requires establishing numerical identity. Numerical identity for persons is particularly difficult given the strict requirements of Leibniz's Law, which stipulates that two things are identical if and only if they have all the same properties. For persons, identity over time requires that a person at two different stages of her life, for example, at age 6 and then at age 15, have all the same properties. It is obvious, then, that in most cases this type of identity will not hold because people change over time. To accommodate this problem, criteria for personal identity over time narrow the scope to only those properties that are necessary and sufficient to establish the survival of a person as the same one over time.

The psychological criterion of identity, often referred to as the Lockean criterion, relies on the continuity of a person's mental features to establish identity over time (DeGrazia 2005, 16). A person at one stage is identical to a person at another stage if and only if they both have the same psychological characteristics. The psychological criterion implies that large changes in personality, including significant shifts in values, preferences, and long-term life plans, could signal a loss of personal identity (Parfit 1984; Perry 1972; Rorty 1976; Shoemaker 1970). The psychological criterion presupposes personhood as essential for the maintenance of identity over time. In other words, identity across time persists only if personhood persists.

The traditional physical criterion establishes a relationship of identity between the person and her body, where the body excludes the brain (Perry and Bratman 1999, 410–416). This version of the bodily criterion excludes the brain because it defines identity in contrast to psychological criteria of identity, and the brain is assumed to realize human psychology. The following thought experiment is a test for the intuitive acceptability of the bodily criterion. Imagine an accident in which two people are injured: A's body remains intact while B's brain is all that is left. B's healthy brain is transplanted into A's healthy body. Which person survives? Most people presented with this thought experiment would think that B survives. This conclusion suggests that the bodily criterion should be considered inadequate (pp. 410–416).

A more contemporary version of the physical criterion, and one impervious to the aforementioned criticism, is the biological criterion, which establishes identity between various stages of the same biological animal (DeGrazia 2005). On this criterion we are human animals persisting through the various stages of development of the body from birth through old age, including the various stages of brain development. According to this view, the body includes the brain. Based on the biological criterion, one can distinguish between the maintenance of personhood over time and the persistence of numerical identity. According to DeGrazia, "there was a time when we who are now persons were not persons (namely, before the human animal developed the capacities that constitute personhood), and there must be a time in the future when we are no longer persons (say, if severe dementia reduces us to barely sentient beings)" (p. 48)

The Human Microbiome and Numerical Identity

The criteria we have described in the previous section are meant to establish the necessary and sufficient elements for personal identity over time. A successful criterion explains how the same person survives across various stages of change. Because both the psychological and the biological criteria aim to establish only what is necessary and sufficient for survival, they are not meant to elucidate all aspects of personhood, nor are they meant to explain the aspects of what it is to be a person that might be important to people.

Most people have some list of features they believe constitute their core self. There is interpersonal variation when it comes to the features people select as the most descriptive of their own personhood. The criteria we discussed in the previous section are not meant to capture any of those conceptions in particular. Rather, they are meant simply to explain how persistence of numerical identity is possible as people change over time, that is, how people can be one and the same in some respects despite changing over time. Both the bodily and the psychological criteria assume an independent way of defining either the human body or core human psychology, without taking into account each person's idea of her core self. Therefore, it seems unlikely that individual conceptual shifts pressed upon us by advances in our knowledge of the human microbiome will change our numerical criteria for personal identity. In what follows, we shall present further arguments for why numerical criteria of personal identity will likely remain unaffected by the HMP.

As we have described it, DeGrazia's bodily criterion for numerical identity is defined in terms of the survival of a human animal. As long as the body, defined as the member of a particular biological kind, persists, the individual survives. The concept of survival in this case includes the persistence of identity: the individual survives change if she remains one and the same individual. For example, a person who has a body part surgically removed or amputated survives that change because her identity remains unaffected. The question is whether the bodily criterion should include the microbiome, and condition survival of an individual on the persistence of the microbiome. Dennett (1991) argues that the proclivity to draw the line between "me" and "the rest of the world" is an expression of the biological self possessed even by microorganisms, and the biological self needs to make a distinction between itself and the environment for protection. This seems to indicate that the survival of the human animal cannot be conditioned on the persistence of the human microbiome. Dennett further argues, however, that even this very basic, biological self is an abstraction, and that the distinction between "us" and "them" does not capture a real boundary in nature (p. 414). The relationship between the human body and its microbiome illustrates this lack of concrete boundary between the body and its environment. Dennett explains, "Within the walls of the human body are many, many interlopers, ranging from bacteria and viruses through microscopic mites, but some of them... are... essential team members in our quest for self-preservation" (p. 414).

We could think of the human as a superorganism comprised of the human body plus the collection of microbes that inhabit the human body. Opting for that different physical criterion of identity, the survival of the superorganism would include the persistence of some aspects of the microbiome. Clearly the human body and the microbiome stand in a symbiotic relationship, and there are many such examples in nature, but those who advance the claim that the microbiome and the human body form a superorganism seem to have something more in mind. One version of the view is that the human–microbial hybrid is comprised of two suborganisms. The view cannot be correct, as the microbiome does not meet the conditions for being an organism: it does not reproduce itself, and it is not composed of interconnected parts such that a cause that affects one part has (roughly) predicable effects in their other parts. For example, if the microbiome in the gut is wiped out, this will not affect the microbiome in the mouth. Even when microbiomic bacteria form a biofilm, the biofilm is not itself an organism, but a kind of structure or scaffold. This lack of organic relation between the parts of the microbiome, or between one segment of a biofilm and a nonadjoining segment, undercuts the case that the microbiome is itself an organism.

Perhaps each regional microbiome, such as the gut microbiome, the vaginal microbiome, and the microbiome of the oral cavity, could be categorized as a distinct organism. Each of those seems to have some of the characteristics of an organism. The microbiome in each of those cavities reproduces itself and maintains a relatively stable microbiome for each individual human. Moreover, the microorganisms constitutive of each regional microbiome are interconnected in terms of their having causal effects on one another. When antibiotics eliminate a certain type of bacteria, that change in environment may facilitate the overgrowth of other organisms. For example, the use of antibiotics can promote the overgrowth of yeast in the vaginal microbiome.

Depending on how we define the individual microbial constituents of the microbiome, the superorganism could be constituted by the human animal plus each regional microbiome, or as the amalgam of the human and the totality of all of the individual microbes. If the microbiome is not an organism per se, perhaps we can conceive of it as a bodily organ. Here the failure of the microbiome to reproduce itself is not an issue; organs like hearts don't give birth to baby hearts. One might complain that my microbiome cannot qualify as one of my bodily organs because it does not contain my DNA. But that verdict seems too strict, since a heart transplanted into me will be one of my organs, as long as it works in concert with the rest of my body.

Even if the microbiome were construed as an organ in the human body, people would not be inclined to associate the persistence of the human animal with the survival of this particular organ. There are times during human life, right after birth, for example, when the human does not yet have a microbiome (Mshvildadze and Neu 2010). So incorporating the microbiome into the biological criterion for human identity over time would complicate our understanding of being human and classify fetuses as members of a different biological kind than other humans.

Returning to the psychological criterion of identity, the changes in our microbiome are unlikely to signal change in our identity, our views about who we are. The realization that 90% of the DNA on our body is not human might make a significant difference in how people see themselves in relation to the environment, how they approach the world, and how they think of microbes. The psychological criterion, however, does not reflect subjective and individual changes in human psychology because it, like the biological criterion, is meant only to establish the necessary and sufficient conditions for explaining how a person remains the same person while changing over time. Any such general psychological criterion that explains survival over time is unlikely to include conceptions of the human microbiome.

To make this point more salient, let us describe in a bit more detail the psychological criterion of personal identity. As we stated earlier, a psychological account of the survival of a particular person over time is associated with the survival of that person's core psychology. On this view, the survival of the body is not essential to the survival of the person. Survival of persons is based on the continuity of their psychology. On this view, people could survive their bodies or change bodies, and numerical identity could still be maintained. For psychological criteria that identified human psychology with the immaterial soul, the survival of the person trailed the survival of the soul, because the soul could survive without the body. Psychological views that do not countenance the existence of the immaterial soul base the possibility of survival over time of a person's psychology on a bit of science fiction. Imagine a time when scientific advances enable us to copy the relevant aspects of human psychology, for example, one's memories, beliefs, and values, onto a memory chip.

We could conceive of people then surviving without their physical bodies. If the psychological criterion enables people to survive their body, and the human microbiome is a feature of the human body, changing conceptions of the human to include the microbiome would not result in changes in conceptions of personal identity.

There are views, however, that do not draw a clear line between the body and human psychology. Human psychology is sometimes treated as continuous with neurophysiology if it is assumed that neurophysiology can explain human psychology. For example, one could argue that proprioception is a fundamental aspect of self-consciousness (Bermudez 2001). If proprioception—which provides information about the location of our body and the location of our body parts in relation to each other—is needed for a conscious sense of self, our psychological identity is dependent on characteristics of our body. Information about the features of our body, on this view, could be constitutive of our sense of self in a very basic way. Put yet another way, the self is embodied.

Nevertheless, even a neurophysiological view of self-consciousness could not be used to argue that facts about the human microbiome should be part of criteria for numerical identity. For one, there is no known mechanism in the body, similar to proprioception, that could generate information about an individual's microbiome in a way that could be constitutive of a sense of self. The immune system certainly interacts with the microbiome, and to do that it must in some sense respond to the microbiome. The immune system must be able to distinguish between the friends and foes of the body to keep humans healthy. But it is not at all clear how the function of the immune system could contribute to a sense of self. Furthermore, even if we could detect features of our microbiome in some yet unknown way—perhaps our vision could be enhanced to detect the microbes on our skin—the same objections that apply to incorporating the microbiome into a physical criterion of numeral identity are pertinent here as well. Just like parts of the body, such as limbs or particular organs that are not necessary for the survival of an individual, a person's body can survive even significant changes in the microbiome, as well as exist for a time without a microbiome. Hence, the human microbiome could not be part of the bodily criterion of identity.

The Human Microbiome and Conceptions of Self

Our review of the different criteria of personal identity showed that the discovery of facts about the human microbiome would not influence the debates about personal identity over time. All of the criteria thus far described could be categorized as third-person criteria, criteria that are distinct from personal conceptions of the self. Third-person criteria, whether psychological or biological, do not rely on features of the self that are important to individual people. Rather, those devising such criteria select particular features of human psychology or biology that are the most likely to remain the same over time and could be used to establish the continuity of personal identity over time. To the extent that they are useful in resolving troubling questions about how we change while remaining the same person, they apply to all people in the same way and in virtue of the same properties, either psychological ones or biological ones. For example, if we had a successful criterion of personal identity, a physician could establish that a person before and after Alzheimer's is one and the same person, even if the person in question currently has no real sense of self. In this way personal identity and conceptions of self diverge. While we might be able to establish a criterion of personal identity without consulting actual people about their sense of self, we cannot really discuss notions of self-conception without relying on the features people deem important for their identity. In this section, we move on to discuss the impact of the human microbiome on those individual conceptions of self.

DeGrazia (2005) describes our individual conceptions of self as constructions of narrative identities that can answer the question of "Who am I?" Each of us has a mental autobiography, which provides an answer to the question of identity. Dennett (1991) formulates the purpose of our narrative selves thusly: "Our fundamental tactic of self-protection, self-control, and self-definition is...telling stories, and more particularly concocting and controlling the story we tell others—and ourselves—about who we are" (p. 418). Our self-conception may depend on our core values, beliefs, personal interests, and characteristics. And there are individual differences in the constitution of our self-conception; people prioritize personal features differently. Some define themselves by their profession, gender, nationality, educational level, and so forth. Furthermore, people's complex conceptions of who they are often guide individuals' future choices. Self-conceptions provide a sense of consistency over time, because each person has the sense that she is acting in accord with who she is. It seems possible that individual conceptions of self could be affected by the HMP and what we learn about the relationship between the human body and its microbiome.

Unlike the philosophical conceptions of personal identity, self-conceptions can be variable and subjective. Each individual's ideas of personal identity should not be taken to specify essential properties either for the individual herself or for anyone else. Unlike third-person criteria, first-person self-concepts need not capture properties that either are objectively true of the person or persist over time. In fact, there could be significant differences in how I think of myself and how others see me.

Furthermore, the traits I find most significant to my self-concept might not overlap with general characteristics that establish identity over time or traits that might define the personhood of all people. Although conceptions of self-identity might superficially resemble psychological criteria for personal identity, they need not accomplish the same tasks as a numerical criterion of identity over time. When generating a conception of self-identity, we need only to formulate a personal autobiography. Moreover, the importance ascribed to features of the self is entirely individual. A person may choose her favorite characteristics and establish a hierarchy between them in any way she pleases. A person might list her values, preferences, and personality traits, and she might consider those features important to her narrative identity, even think of them as constitutive of her core self. Such personal impressions do not establish a more generally applicable criterion for personal identity, and one person's notion of personal identity does not restrict the conceptions of self for others. In other words, personal conceptions of self-identity are not normative. They do not establish philosophical standards for what *should* be true of any individual.

Each of us could prioritize the elements of our core self-identity in a very different way. Some might prioritize their professional identity over any other kind of interest, value, or personality feature. Others might define themselves in terms of the role they have in their family, as daughter, mother, or favorite aunt. One might have inaccurate conceptions of oneself. For example, a person might think of herself as having a great sense of humor, and perhaps make elaborate efforts to be humorous, even if such an impression is not shared by anybody else. Because we each have an individually defined conception of self-identity, accuracy is not a requirement, although it might be plausible to expect that over time people might revise their inaccurate self-conceptions.

Individually conceived identity need not capture what are often thought of as important features of personal identity. A person's core identity need not include traits that others consider constitutive, such as religion, gender, sexual orientation, nationality, profession, and so on. One could prioritize relatively unimportant personal features such as being thrifty or being a good dresser as part of one's core identity and make efforts to act in accordance with that self-image.

It seems quite obvious that people change their conceptions of self over time. In an often-cited example from his book *Reasons and Persons*, Derek Parfit (1984) describes a young Russian with strong socialist ideals who wishes to give land to the peasants after he dies. He suspects, however, that with age he might change his mind and become less willing to share his wealth and thus asks his wife to never let him revise his will, which endows the peasants with his land (p. 327). For a numerical criterion of personal identity, the change in political views the young Russian foresees might signal a loss of personal identity, and any criterion aimed at establishing personal identity over time would have to accommodate such large shifts in personal psychology. Personal conceptions of self can tolerate large shifts in personal psychology because the numerical identity of a person across time is more or less taken for granted.

As we change over time, grow older, or learn new things, the relative importance of our core traits might change. These changes in identity can be traced back to a variety of sources. A person may change her social status, obtain a college degree, become part of a family unit, or even join a religious cult. Thus, if attending college, joining a religion, or even reading a good book could affect a person's sense of self, there is no reason to presume that scientific discovery could not have a similar effect on conceptions of personal identity.

It might be the case that conceptual shifts resulting from the research on the human microbiome come to influence how some people think of their identity. In what follows, we present some examples of how that might happen. These examples are meant to illustrate how this process of conceptual change could occur, but they are not meant as committed predictions that the changes will occur as described.

Some studies have shown that the salivary microbiome might differ across countries (Grice et al. 2009). People who learn this could assimilate particular features of their microbiome into their sense of national identity. Changes in my microbiome could affect my ideas about who I am, racially or nationally, or they might enhance or diminish my sense of regional identity. Supposing counterfactually that microbiota do not differ across regions, one could see how such a scientific fact could diminish a sense of national identity and perhaps promote a sense of self that is cosmopolitan and not based on country or region of origin.

Changes in the microbiome could signal changes in self-concept in other ways as well. Those who are particularly fearful of bacterial contamination might change their views after hearing that some bacteria in the human microbiome promote health. Presumably, persons who are fearful of bacterial contamination are worried about the pathogenic character of bacteria. If the relationship between bacteria and the human body is reconceptualized, each person's stance toward bacterial contamination might change. Hence, facts about how bacteria relate to the human body can contribute to the development and prominence of certain psychological characteristics in humans, which in turn contribute to conceptions of self.

The knowledge that each of our microbiota might be unique and could be used to identify us might encourage identification with one's microbiome. We could come to incorporate our unique microbial mark as a unique expression of the self in similar ways as fingerprints. Microbiota might reflect eating habits, past treatment with antibiotics, and travel to different parts of the world. In turn, that difference could affect how we conceive of ourselves, and how we conceive of ourselves in relation to our environment. The discovery of the human microbiome might eliminate the dichotomy we established between ourselves and our environment and could prompt us to realize that our identity is in part determined by our environment. Even more importantly, it might help us to understand how our very proximate environment, comprised of our microbiome, can be essential to good health.

An additional way in which the HMP might influence our conceptions of self is by providing supportive evidence for the notions of a human as a superorganism. Battin et al. (2009, 77) introduce the notion of the "way-station" self and argue that the "individual" *should* be seen as "a large organism carrying and inhabited by a host of smaller ones that move easily from their habitat in one 'person' to another." Their argument calls for a change in perspective so that each person comes to see herself within a particular social and biological context, where our identity may change from being a victim of disease to being a vector of disease. Their strongest claim is that individuals are not distinct from their environment; rather, people *are* their microbial environment (p. 81).

This view could have an impact on philosophical conceptions of personhood. For some philosophers the notion of person is a morally laden concept. Philosopher Immanuel Kant and many others discuss ethics in terms of accountability, rights and duties, human dignity, and rational agency. They see a person as a moral agent. From their perspective, the word "person" confers an aura of dignity that no mere parcel of matter or physical specimen can claim. To speak of something as a person is to refer to an entity that is worthy of respect, endowed with the capacity for freedom, autonomy, self-consciousness, intelligence, imagination, communication—the cluster of normative attributes may vary with the philosopher. For Kant, there is a morally fundamental distinction between persons and things. The conception proposed by Battin et al. of person as a way-station might change how we conceive of our personal responsibilities and extend those to the effects of our microbiome. It might also limit the notion of our autonomy because microbes move from person to person freely, thereby limiting our ability to control our impact on the health of those in our immediate environment.

Despite these conceptual changes, it seems doubtful that people will come to think of their microbiome as a defining feature of their self-concept that is as important as being a member of a profession or being a parent. It might be that at a certain time in a person's life, perhaps during an illness, one might focus more on the health of her microbiome. In those situations a person might, for a time, think of herself as being defined by her microbiome. She might invest significant time and effort in improving her health by improving the health of her microbiome, just as people now spend significant time and energy in improving the health of their bodies with exercise and nutrition. Indeed, some of the very same activities might be reclassified as ways of improving one's microbiome rather than one's body.

Even if the microbiome does not become a defining feature of the self, knowledge about the relationship between our bodies and our microbiome could influence people's willingness to participate in research. As the microbiome becomes incorporated, even marginally, into people's conception of self, they might become more drawn toward participating in research on the human microbiome. Willingness to participate in research might increase especially if it is shown that there is a relationship between a healthy body and a healthy microbiome.

The Impact of Science on Commonsense Conceptions of Self

Although we are far from thinking of ourselves as essentially defined by our microbial environment, there is some evidence to show that we are beginning to incorporate knowledge about the human microbiome into our sense of self. As we already mentioned, commonsense conceptions of personal identity and personhood are variable and may be influenced by personal experiences, acquisition of new knowledge, or appropriation of a current zeitgeist. Commonsense conceptions of personhood are often affected by scientific discoveries as well. Nerlich and Hellsten (2009) trace the impact of the HMP on our use of metaphor when describing the symbiotic relationship between "us and them." They argue that along with changes in our use of language, our conceptions of self will be influenced. They cite the following from recent press reports:

> "So you think you are the self-reliant type. A rugged individualist.... You'd be nothing without the trillions of microbial minions toiling in your large intestine...."
>
> "You are suffering from a serious crisis of identity. Scientists believe you are not entirely human. In fact, it is time to stop thinking of yourself as an individual..."
>
> "... [E]ach of us, sick and healthy, could start speaking in the first person plural, not with the editorial, but the biological 'we.'... [W]e are what lives within us."
>
> "Microbes are a part of who we are."

Although it might be difficult to evaluate the actual impact of such general proclamations, we could make a prediction based on conceptual shifts stemming from past scientific discoveries. The HMP could have a similar influence. One such conceptual shift in self-image was caused by the scientific discovery that mental activity is localized in the brain. This knowledge is now part of commonsense psychology, although it has not always been so. According to the ancient Greek philosopher Aristotle, sense perception, a type of thinking, is associated with movement of blood in the body and functioning of the heart (Ackrill 1987, 212–216). Aristotle described the correlation between movement of blood and emotions. Anger, for Aristotle, corresponded to "the boiling of the blood and hot stuff around the heart" (p. 163), and spontaneous movements in the sense organs as well as the movement of blood during sleep caused dreaming.

If we contrast Aristotle's view with the current way we speak of mental states, one can clearly see a change in how human psychology is explained and described. This change is present not only in scientific circles but in everyday life. The association between "chemical imbalances" in the brain and psychological moods has become a platitude of everyday conversation. The existence of neurotransmitters in the brain and their association with conditions like depression and other mood disorders has permeated common sense. People who exercise often report that they have a high after physical activity and attribute the changes in mood to increased production of serotonin. The association between "damage to brain tissue" and loss of cognitive abilities is widely known, and most people associate Alzheimer's disease with changes in the brain. These examples show that neuroscience has had a significant impact on how we speak of our psychological states, how we interpret our own behavior and that of others, and how we explain and predict behavior.

Although in everyday parlance one still hears expressions that refer to the heart as the locus of our emotions, those references are mostly metaphorical. Poetry and songs refer to heartbreak, but emotional pain is no longer associated with the physiological functioning of the heart. The claim being made is not, however, that one could not find a person, or even groups of people, who associates damage to the heart with emotional pain. The point here is that, for the most part, our current commonsense understanding of human psychology no longer features that association. In addition, our current association between brain function and human psychology is a result of scientific inquiry.

Our commonsense understanding of psychology also reflects a rejection of the Cartesian view of mental states. On Descartes' view, all mental states are conscious states. The popularization of Freud's writings has affected how we see the relationship between consciousness and mental states. Freud's theory has shaped our commonsense views by introducing the idea of unconscious mental states into our everyday parlance. Nowadays people often refer to unconscious thoughts and motives and accuse each other of "Freudian slips." There is some controversy with regard to whether Freudian theories about human psychology are scientific, but that particular debate is not relevant here. The argument is not that only good science becomes incorporated into common sense; it is that scientific knowledge, broadly construed, has shaped common sense.

The HMP has also influenced our conceptions of ourselves. Mapping the human genome and the association of many human traits with DNA has caused waves of conceptual shifts. We now see our physical traits, inclinations, and medical ailments in light of our genes. The discovery that there is a genetic component to psychiatric diseases such as depression and schizophrenia has influenced our view of the boundaries of human agency and identity and limited our estimation of the power of free will.

The aforementioned examples illustrate how the introduction of scientific knowledge into everyday life can transform our concepts and affect our narrative identities. Conceptual shifts are not mere acknowledgment of additional facts; they entail changes in how we perceive and interpret our experiences. To the extent that an understanding of the relationship between humans and their microbiome permeates common sense, we may come to *speak* of ourselves as superorganisms and see ourselves as conglomerate beings. Such conceptual shifts could change how we behave. They might prompt us to promote the health of our microbiome. They might also encourage us to take more responsibility in protecting others from contagion and encourage us to extend greater effort on conserving the environment.

As Battin et al. (2009) advocate, we ought to think of ourselves as way-stations. Thus, our moral responsibility extends to cover our microbiome and all the ways in which it might have an impact on our environment and the individuals with whom we come into contact. Such a view would be an extension of how we currently conceive of our responsibilities to our environment. For example, if one has an infectious disease, one is expected to avoid spreading it to others. The acknowledgment of the relationship between our body and our microbiome might produce other similar responsibilities to promote our own health. To an extent, one could even advocate that we have a responsibility to be well informed about the human microbiome. In the context of health care and biomedical research, thinking of ourselves as being comprised of human and microbial parts might incline us to think more about promoting societal benefit through participation in research.

Moving beyond self-conceptions of identity, the HMP could also impact how both scientists and physicians think about the promotion of human health. Given that a healthy human organism depends in some part on its bacterial inhabitants, they could come to see a human as a superorganism—a human–bacterial hybrid. Such a conceptual shift will affect the direction of biomedical research and promote nurturing a positive symbiotic relationship between us and our microbiome. As a result, physicians might shift more of their attention to the influence of the environment on health and begin to treat not just the person but the superorganism.

Research on the microbiome could also impact national, race, and gender identity because the HMP might yield data about particular populations. For example, people who live in the same country, or region, might have similar microbiomes. In such cases features of the microbiome could come to be associated with national identity, in the same manner that other group traits sometimes are. Differences in microbial samples could reveal that we have more in common with our neighbors than with our geographically distant relatives or others of similar national origin. In countries where there is significant ethnic diversity, characteristics found in microbial samples could help to redraw communities based on shared environment rather than shared ethnicity.

Given the effects of different combinations of microbiota on health, such discoveries could be either positive or stigmatizing. Therefore, it will be important to gauge the impact of information about the human microbiome and decide on appropriate ways of communicating the information to the public.

It will also be important to appreciate the influence of scientific language on everyday life. We explained how scientific facts seep into commonsense conceptions of self. But the influence of scientific knowledge extends much beyond personal conceptions of self to issues much beyond the scope of this chapter. The rapid development of science and technology and the increasing specialization in scientific fields encourage scientists to develop a technical language that isolates those with scientific expertise from everybody else. Science, however, is democratic in its influence on human life. Thus, it is the responsibility of scientific experts to promulgate scientific results in language that enables both individuals and policymakers to understand the information and to use it in making informed decisions (Lewontin 2002).

Conclusion

In this chapter we considered how personal identity has been treated in the philosophical literature and how personal identity is defined in everyday parlance. We reviewed two of the most prominent criteria for personal identity over time, the psychological criterion and the bodily criterion. We also distinguished between criteria for personal identity and individual conceptions of self and argued that they are distinct notions. Philosophical conceptions of personal identity will not be affected by the research conducted under the auspices of the HMP. We argue, however, that personal concepts of identity, or self-conceptions, are likely to be influenced by the facts about the human microbiome. Because scientific facts have shaped conceptions of self in the past, facts about the human microbiome, and how it interacts with the human body, might influence our ideas of personal identity. National identity, for example, could be solidified through the discovery of regional differences in microbiome. Similarly, individuals fearful of bacteria might change their minds after they become apprised of the fact that bacteria are not always pathogenic. Some might even adopt the notion that a person is a human and bacterial hybrid, a superorganism.

Policy Recommendations

- How the research on the human microbiome is publicized and how it is ultimately incorporated into people's sense of self will depend on the way in which facts about the human microbiome are communicated to the public. We recommend that scientists working on the Human Microbiome Project take the lead in disseminating the results of their research in language that is accessible to the public to ensure that the information is communicated accurately.

- As we learn more about the connection between our health and our microbiome, our obligations to maintain the health of our microbiome might increase. In addition, because each individual's microbiome can, through contamination, have a negative effect on the health of those in our proximate environment, individuals will be responsible for maintaining the health of their microbiome.

References

Ackrill, J. L. 1987. *A New Aristotle Reader*. Princeton University Press.

Battin, M.P., L. P. Francis, J. A. Jacobson, and C. B. Smith. 2009. *The Patient as Victim and Vector: Ethics and Infectious Disease.* New York: Oxford University Press.

Bermudez, J. L. 2001. Nonconceptual self-consciousness and cognitive science. *Synthese* 129: 129–149.

DeGrazia, David. 2005. *Human Identity and Bioethics.* Cambridge University Press.

Dennett, Daniel C. 1991. *Consciousness Explained.* Back Bay Books.

Grice, Elizabeth, et al. 2009. Topographical and temporal diversity of the human skin microbiome. *Science* 324(5931): 1190–1192.

Lewontin, Richard C. 2002. The politics of science. *The New York Review of Books,* May 9.

Mshvildadze, Maka, and Josef Neu. 2010. The infant intestinal microbiome: friend or foe? *Early Human Development* 86(July): S67–S71.

Nerlich, Brigitte, and Lina Hellsten. 2009. Beyond the human genome: microbes, metaphors and what it means to be human in an interconnected post-genomic world. *New Genetics and Society* 28(1): 19–36.

Parfit, Derek. 1984. *Reasons and Persons.* Oxford: Clarendon.

Perry, John. 1972. Can the self divide? *Journal of Philosophy* 69(16): 463–488.

Perry, John, and Michael Bratman (eds.). 1999. A dialogue on personal identity and immortality. In *Introduction to Philosophy: Classical and Contemporary Readings* (3rd ed., pp. 396–416). Oxford: Oxford University Press.

Rorty, Amelie, ed. 1976. *The Identities of Persons.* Berkley: University of California Press.

Shoemaker, Sydney. 1970. Persons and their past. *American Philosophical Quarterly* 7: 269–285.

3

Property and Research on the Human Microbiome

Abraham P. Schwab, Mary Ann Baily, Joseph Goldfarb, Kurt Hirschhorn, Rosamond Rhodes, Brett Trusko

Four Controversial Cases

CASE 1

Ananda Mohan Chakrabarty tried to patent a bacterium that could break down crude oil and thus help in cleaning up oil spills (Robinson and Medlock 2005). His application was rejected on the grounds that living organisms could not be patented. He sued the patent office (*Diamond v. Chakrabarty*). The court ruled that his bacterium was patentable because it was man-made, since he had modified the DNA of an existing organism to create it.

CASE 2

Myriad Genetics, a biotech company, held patents for two genes related to breast and ovarian cancer and a diagnostic test for the genes' presence. The patents enabled the company to block development of alternative tests and insist on being the sole source of its own test. Samples for testing must be sent to the company and a large fee paid. In 2009, several groups joined forces to sue the patent office (*Association for Molecular Pathology et al. v. U.S. Patent and Trademark Office, Myriad Genetics, et al.*), claiming that genes are not patentable because they are natural entities. The plaintiffs won initially but lost in the Court of Appeals for the Federal Circuit (CAFC). They appealed to the Supreme Court, which vacated the CAFC ruling and returned the case to that court for reconsideration in light of the high court's ruling in another case (*Mayo Collaborative Services vs. Prometheus Labs, Inc.*) that companies cannot patent observations about a natural phenomenon (Pollack 2012). The CAFC upheld the patents again and the plaintiffs appealed the ruling back to the Supreme Court. As of November 2012, the final outcome was still undetermined (Saunders 2012).

CASE 3

A best-selling book (Skloot 2010) tells the story of the HeLa "immortal" cell line, a valuable tool in medical research. The original cells came from Henrietta Lacks, a poor black woman who was being treated for cervical cancer; they were taken for research purposes without her knowledge and consent. Critics have noted that many others benefited, financially and in other ways, from "Henrietta's cells," while her family remained in dire poverty, unable to afford health insurance.

CASE 4

John Moore's spleen was removed during his treatment for leukemia (Lawnix 2012). Without Moore's knowledge, his doctor used tissue from the spleen to create a cell line, patent it, and profit from it. Moore sued the doctor and the others on the patent (*Moore v. Regents of the University of California, et al.*). The suit claimed that by failing to disclose his economic interest in the cells, the doctor breached his duty to provide Moore with information he needed to give informed consent to treatment. It also claimed that by taking and patenting the cells, the doctor violated Moore's ownership rights in them. The final outcome of the litigation was a California Supreme Court decision that agreed with the first claim but rejected the second, saying Moore had no ownership rights in the cells after removal.

Introduction

Embedded in these cases are difficult questions about property in the context of biomedical research. Are cells, genes, tests, research techniques, and research results property that can be owned? What rights does property ownership confer? Can a property owner exercise complete control over who can use the property and for what purposes? What rules should govern the process of making a profit from the property?

Research on the human microbiome will raise similar property issues. The Human Microbiome Project (HMP) is a massive international collaboration that will produce valuable scientific knowledge about the human microbiome and the role it plays in human health.[1] Microbiome researchers will identify new organisms, invent new laboratory processes and techniques, and develop databases that will be used in further studies. Research may also lead to marketable products such as diagnostic tests and foods that improve health by modifying an individual's microbiome. Individuals provide the biological samples that are the material for research. Some are parts and products of the human body, such as blood, skin cells, and saliva; others are samples of the microbiome, the microorganisms that live on or in the body. Policymakers and the public must be prepared for the inevitable questions about what, if any, property rights those who provide samples and those who use the samples to carry out research have in the scientific knowledge produced, the natural entities identified, and the processes, products, and data developed.

In this chapter, we provide a framework for understanding the property-related microbiome issues that are on the horizon. We begin by explaining the concepts of ownership, property, and property rights. Next we discuss how these concepts have been applied in the U.S. research environment. Then we explore some implications for the management of property as research on the human microbiome proceeds.

Ownership and Property in Philosophy

The social contract theories advanced by the philosophers Thomas Hobbes, John Locke, and Jean Jacques Rousseau are widely acknowledged for their role in informing our understanding of the modern liberal democratic state. These theorists took radically different positions on property. Their different views reflect competing positions on property that were fiercely debated in their time and continue as controversial issues today.

"Property" has been defined as a "claim, of an individual . . . to some use or benefit of some thing" (Macpherson, 1985, 87). This definition applies to private property, giving the owner of the property the right to exclude others from use as well as rights to dispose of it or transfer it to another. The definition also applies to common property by giving owners the right not to be excluded by others. In the history of philosophy, since at least the writing of Aristotle, the scope of property has included both sorts of claims: the exclusive use of private property and the accessibility of common property.

---

[1] The article "The NIH Human Microbiome Project" in *Genome Research* describes the origins of the collaboration, the goals and structure of the National Institutes of Health's (NIH's) microbiome project, and the vehicle for international cooperation, the International Human Microbiome Consortium (IHMC; Peterson et al. 2009). The consortium began with 10 member countries and is open to any microbiome research project that is willing to agree to the IHMC's founding principles. Current information about the NIH HMP is available at http://commonfund.nih.gov/hmp/ and http://www.hmpdacc.org/. Information about the IHMC is available at http://www.human-microbiome.org/.

Property includes material property in things. Material property is both those things that a society can allow to be privately owned, such as one's home, one's land, one's hairbrush, and one's car, and those things that a society declares to be for common use, such as public buildings, parks, roads, and reservoir water.

With respect to the human microbiome, a society could declare that the microbiome is material privately owned property that is within or upon one's body or produced by the individual's distinctive environment. Alternatively, a society could treat knowledge about the human microbiome as a communal resource. By employing our existing legal frameworks and creatively developing new structures, we will have to sort out precisely which property rights in the microbiome should be counted as private property and which should be treated in other ways. Long before we knew about the human microbiome, philosophers wrestled with how to understand property and to distinguish private from common property. Their insights can be useful in sorting out the positions on property that may be employed in discussions of how to regard the microbiome.

Philosopher Thomas Hobbes maintained that by nature no one owns anything at all. In *Leviathan* (1651), Hobbes argued that nothing is naturally yours, not even your body. Things belong to you only so long as you can hold onto them. By nature, those who are stronger than you can do whatever they are able to do to your body, even use it as a shield. Claims to property ownership are, therefore, not natural but a matter of convention, created by societies as a way of avoiding civil strife and maintaining social stability. Thus, on his view, property and property rights are entirely social constructions (Rhodes 2009).

For Hobbes, therefore, property rights are nothing like the immutable laws of nature; they are far more arbitrary. They are political tools that can and should be accorded and manipulated in service of achieving and maintaining a stable and flourishing society. Anything that can be exchanged, including human bodies, land, physical things, money, labor, products of labor, products of natural talents, or products of developed skills, is subject to distribution or redistribution to achieve the social goals of peace, meeting basic needs, and sharing the luxuries provided by the society.

According to Hobbes, everyone can acknowledge that self-preservation requires maintaining some control over the necessities of life. Thus, we should each allow others control over some things so long as they allow that same control to us. How far that control extends will be governed by local laws. The extent of the control will be different in different places and at different times. Property rights are local, created by the ordinances of the state, and they may be altered to meet changing circumstances and the exigencies that arise. Hobbes also identified various levels of property claims. Specifically, he claimed that individuals' private property comes in at least three degrees: "his own life and limbs; and in the next degree, (in most men), those that concern conjugal affection; and after them riches and means of living" (Hobbes 1651, chap. 30, p. 382). For Hobbes, whereas one's body and conjugal affection are reserved for exclusive private use, means of living should often be considered as common property.

This Hobbesian view of property suggests that granting property rights to the human microbiome should be aligned with social goals. Given the ongoing disagreements about property rights, and protections of intellectual property and patents in particular, developing policy related to property rights associated with the human microbiome will have to be carefully considered in light of the social goals of our time and place.

In contrast to Hobbes, John Locke viewed the human body as natural property. His most significant discussion of property can be found in his *Two Treatises of Government* (1689). There he offers that, "Every man has a property in his own person. This nobody has a right to, but himself" (p. 111, Ch. V, sec. 27). His view of property derives from his view that the world belongs to all persons in common. Locke separates the world into private property (the body) and common property (everything else). This distinction raises the question of how nonprivate property comes to be distributed as private property, often unequally. Locke's answer is that individuals acquire as property those things with which they have mixed or invested their labor. For example, the words written on this page are the property of the authors, because the words are the products of our labor. Nonetheless, Locke also takes life, liberty, and health to be natural rights of human beings. As such, no amount of labor can give one person the right to deprive another of life, liberty, or health.

As Locke explains, "The reason why men enter into society is the preservation of their property." As a result, the only legitimate taxes are those that ensure property rights. Any other attempt by government to claim a share of an individual's property or redistribute it is a usurpation. Robert Nozick and other contemporary libertarians draw from Locke's views to defend the idea of property as a domain over which the individual maintains absolute authority. For libertarians, individuals have irrevocable property rights in their own bodies and anything with which they mix their labor.

In one respect, the implications of Locke's view of property for the human microbiome turns on whether or not the human microbiome should be considered a person's own body. On the one hand, the microbes that live on us and in us can be considered not to be part of the person. Some of our microbiome can be added, or removed or replaced, without any apparent change in our body. While many people ignore the microbiome, others try to destroy it with antimicrobial products. On the other hand, the microbiome inhabits our bodies and constantly interacts with the individual's physiology and genetic makeup. In that light, the microbiome could count as part of us. Adopting Locke's view would make sorting out whether or not the microbiome "really" is part of us the core issue related to property claims in the human microbiome.

Jean-Jacques Rousseau understood property as a common resource to be shared among the population for the public weal. In direct opposition to Locke's view, Rousseau ([1755] 1992, p. 45) cautioned that:

> The first man who, having enclosed a piece of ground, bethought himself of saying This is mine, and found people simple enough to believe him, was the real founder of civil society. From how many crimes, wars and murders, from how many horrors and misfortunes might not anyone have saved mankind, by pulling up the stakes, or filling up the ditch, and crying to his fellows, "Beware of listening to this impostor; you are undone if you once forget that the fruits of the earth belong to us all, and the earth itself to nobody.

In the *Social Contract* ([1762] 1968), we find Rousseau's most robust discussion of property. Like Hobbes, Rousseau understands the distribution of property to be a matter of social concern. The difference lies in how this concern is addressed. For Rousseau, the validity of any allocation of resources to individuals as property is determined by the general will, that is, whether or not from their natural, innate goodness, self-love, and reason all parties consent to the property allocation in question. If the consent of all to an allocation of a common resource cannot or, in exigent circumstances, could not reasonably be expected to be achieved, then the property claim is illegitimate.

Lack of specificity is the main difficulty with Rousseau's view of property, a problem that also arises in the similar views of Immanuel Kant and John Rawls. Actual consent of all parties is often not achievable or is impractical. As a substitute for actual consent, theorists revert to a standard that conceptually approximates it, that is, a position to which any reasonable person would consent or cannot reasonably withhold consent. Determining the specific boundaries of this possible rational unanimity, or even setting a noncontroversial definition of reasonableness, can be contentious.

Intellectual property (IP) is a relatively new category of property concerned specifically with the use of ideas. IP may be considered either as private property with an owner who has rights to exclusive use or as a common resource. IP may also be viewed as the sort of thing to which people have rights to obtain access. In recent developments we have witnessed IP claims related to microbes falling into each category. For example, manufacturers have identified strains of yeast and secured patents to give them private property rights in them. In contrast, the HMP has required human microbiome genetic information to be posted on open websites, thereby treating it as common property.

IP does not require exclusive use. While only one person at a time can use my fishing pole to catch a fish, any number of people can use the idea of fishing with a hook and line to make their own poles and go fishing. This raises the question, then, why does someone have a right to the exclusive use of ideas? To answer in the affirmative, one must identify what gives rise to this right.

There have been several different attempts to do this. For example, some have used Locke's argument that one's right to a thing as one's property arises from one's labor. Thus, one has a property claim on those ideas or interpretations or expressions that result from one's labor. As with the discussion of material property, however, there is ambiguity here. As noted earlier, intellectual property is not burdened with the necessity of exclusive use. The question, then, for Locke's view is why it would be necessary to limit the use of an idea. One might think of it this way—while the individual who first had the idea (or interpretation, expression, etc.) had to labor to produce the idea, any other individual who comes to understand the idea also has to labor to understand it. What, then, gives rise to the exclusive use for the first individual to use the idea? Again, we can see advantages to adopting a roughly Hobbesian understanding of property and structure for its regulation. Because Hobbes determines the distribution of property according to identifiable social goals, we can provide a meaningful answer to this important question.

Attempts to justify exclusive use of IP can be grouped into two general categories: principled and consequentialist. Principled justifications for granting intellectual property rights tend to rely on claims that an individual is owed exclusive use. For example, using Locke's view as a basis, one would argue that through the effort invested, the individual has produced a property right. Not granting an IP right would unjustly deny the individual what was earned through personal effort. Consequentialist explanations are more prevalent and more straightforward. Basically, consequentialists argue that individuals should own property because this arrangement has identifiable social benefits. Private ownership is an incentive for innovation and creativity. By allowing individuals to have exclusive use for some time, individuals are relieved of worries that they will not derive benefits from their hard work. If society does not protect their exclusive use, individuals will be less likely to invest their effort. They will be worried their efforts will only serve to line the pockets of others, who, without putting in the time and effort that they have, simply use their innovation. The costs are incurred by the innovator. Without assurance that the innovators will reap the profits of their labor, the investment of resources, time, and energy would be unreasonable. Without innovation, society would miss out on the benefits that they could provide.

Consequentialist justifications for granting intellectual property rights to innovators are subject to critical review. That is, there is an expectation that social benefits will accrue: for example, when we compare areas where robust intellectual property rights are in place with places where there are no such rights, there should be a noticeable difference between the innovations produced in these areas.

Ownership and Property in the Social Sciences

This brief review of the history of philosophical discussions of property suggests that a Hobbesian approach may be most helpful for understanding how real societies manage property issues. Studies of property in the social sciences confirm this conclusion. The management of property issues is a significant feature in how societies are organized. Thus, in anthropology, sociology, history, political science, and law, scholars pay close attention to land tenure systems, rules of inheritance, constraints on what can be owned and exchanged, and other property-related social and political structures. These studies reveal that in very simple societies, ownership and property are matters of custom. As a society becomes more complex, customs are incorporated into a legal system that can interpret them and enforce compliance. The establishment and enforcement of property law are a central function of government. In legal systems based on common law, interpretation and refinement of custom have developed over centuries through adjudication of individual cases. Modern governments also proactively create new law to reconcile conflicting ownership claims and to address societal interests related to scientific and technological advances.

Economic analysis of ownership and property pulls together the perspectives of law and the other social sciences by characterizing legal rights as a mechanism a society uses to achieve social goals (Barzel 1997). In economics, the term *property right* has two distinct meanings. The traditional meaning is the right to property that is recognized and enforced, in part, by the state. Property rights are legal rights that serve as the constraints within which economic agents must function. The newer meaning identifies property rights as an economic concept distinct from the legal right to property. An economic property right is defined as a person's expected ability to consume a good or the services of an asset directly, or indirectly through exchange. In other words, an economic property right is the de facto ability to use or trade something.

Property rights in the economic sense are neither constant nor absolute. They depend on the actions of the people who hold them and the actions of others: the right-holders' own efforts to protect their rights, the attempts of other people to capture or protect the rights, and the government's efforts to protect the rights through police and court intervention. The economic definition implies that a person's economic property rights over a commodity that is subject to restrictions on its exchange or is prone to theft are diminished. Economic property rights are thus clearly related to but not identical to legal property rights; for example, "squatters' rights" exist, but they have to be self-enforced.

The concept of *transaction costs* plays a critical role in the economic analysis of property rights. The term refers to the costs associated with engaging in economic transactions, such as the costs of obtaining information about the attributes of a good to be exchanged and the trading partners' respective terms, the costs of writing contracts to establish which attributes are changing hands, and the costs of carrying out the exchange. Defining economic property rights in terms of the de facto ability to use or trade attributes of a good provides useful insights into the economic role that legal property rights play. Without a legal right, the owner bears the burden of enforcement (the squatter's rights case). If the owner has a legal right, his or her claim is easier to enforce and taxpayers bear some of the enforcement cost. Moreover, the existence of clearly defined legal rights often reduces the other transaction costs associated with property use and exchange.

In sum, the social science literature on property ownership shows that societies assign and restrict property rights as a matter of social policy. Property rights vary significantly over different domains within each society, for similar domains across societies, and over time. Analyses informed by philosophy and the social sciences can help us understand how different societies conceptualize property rights and how factors such as a society's cultural values, natural resource endowments, and level of economic development produce the observed variations in the treatment of property. Such analyses also show that in most societies, including U.S. society, the members often have conflicting views on property rights and how they should be managed. This results in inconsistencies and contradictions in the rules governing ownership and property, especially the rules that apply to intellectual property and to human body parts and products.

Ownership and Property in the Context of Biomedical Research

Research is designed to advance knowledge. The results of research can range from new information about fundamental physical, chemical, and biological processes that expands our understanding of the universe we inhabit (basic research) to practical applications of that understanding to produce things that are useful in everyday life (applied research). Traditionally, most *basic* research in the natural sciences in the United States was funded by the government or by private philanthropists, and much of it was done within academic institutions by salaried scholars engaged in both research and teaching. Within this framework, research results were common property. Researchers were honored for producing knowledge that society valued, but they didn't own the knowledge or control its use. Dissemination of research results was the norm, and the incentive to do good research was the inherent interest of the work, bolstered by the prestige attached to its originality and social importance.

Basic research in the natural sciences often provides the foundation for the development of products and processes with significant economic value (Nelson 1959). In the biomedical field, for example, it can be the basis for new drugs and new medical devices. Traditionally, for-profit corporations took on the *applied* research and development that led to such products and processes and obtained IP protection giving them exclusive marketing rights. Although research universities contributed significantly to industrial innovation in other ways, until the mid-1970s, they were generally reluctant to become directly involved in patenting and licensing activities.

Since that time, the research environment has changed dramatically. The line between basic and applied research has always been fuzzy, but now there is much less separation between the two. Moreover, researchers are less committed to the idea that research results are to be shared as common property. Today academic researchers, universities, and other nonprofit research organizations are frequently involved in entrepreneurial, for-profit activities designed to enable them to obtain direct financial benefit from their research results. A critical event in this shift was the passage in 1980 of the federal Bayh-Dole Act. The act modified rules and incentives for federally funded researchers and their institutions in ways that encouraged them to seek patent protection and develop inventions related to their research; equally important, it changed social norms about the appropriateness of university involvement in the business of patents and licensing (Boettiger and Bennett 2006; Rai and Eisenberg 2003; Sampat 2006).

The shift has been especially evident in biomedical research. Researchers and research organizations not only have been claiming intellectual property rights in end products, such as drugs, diagnostic tests, and devices used in medical treatment, but also have been patenting products and processes developed for use in the research process itself. This forces other researchers to obtain permission and sometimes pay royalties to pursue work on the frontiers of knowledge. In fact, there is growing concern that things have gone too far and researchers are obtaining IP rights in basic science, endangering the research process itself. Commentators argue for the social value of the *science commons,* defined as knowledge that is widely accessible at low or no cost. For example, Robert Cook-Deegan calls the science commons "a uniquely important input to scientific advance and cumulative technological innovation" that "diverges from proprietary science primarily in being open and being very broadly available" (Cook-Deegan 2007, 133). There are calls for new policies to push back against the trend of claiming private property rights in research results—for example, revisions of the Bayh-Dole legislation (Colianai and Cook-Deegan 2009; Cook-Deegan 2007; Cook-Deegan and Dedeurwaerdere 2006; Eisenberg and Nelson 2002; Nelson 2004).

As researchers and their home institutions have sought to profit directly from their research activities, controversy has also arisen over the use of human biological samples in research that results in products and processes with economic value. People who have provided samples without compensation feel injured and used when they see others making large profits from the samples. Some samples were collected specifically for research, whereas others were collected for clinical care purposes and leftovers used for research instead of being discarded. Do people have property rights in biological samples collected from their bodies? Does the answer to that question provide useful guidance in settling the compensation and consent issues that arise when the samples are used in research?

This chapter considers the arguments for and against an expansive system of IP protection in biomedical research and explores the questions raised earlier about property rights in human biological samples. The key property issues relate to the distinction between material and intellectual property, the distinction between common and private property, and the special treatment in property law of the human body, its parts, and its products. We explain the current understanding of these concepts and apply them in the context of the human microbiome.

Current Property Structures and Research on the Human Microbiome

As already noted, *material* property refers to tangible things, and *private* property is property that is set aside for the exclusive use of one individual or a group of individuals. In practical terms, a privately owned piece of material property can be understood as a bundle of rights relating to the attributes of a thing. A specific thing—for example, a factory building—can have multiple attributes, which can be positive or negative; the factory may produce a useful product but also pollute the air. The *property rights* to the different attributes can be held by different people and traded separately from one another. For example, an apartment building can be owned by a landlord, who trades the right to occupy the building for a specific time period to a tenant in exchange for rent. The tenant may or may not have the right to sublet the rented space to someone else. The landlord may own the building but lease the land it is on; if he owns the land, he may or may not have property rights to the minerals underground or the air space over the property. In a rural area, the owner may be able to rent the right to graze animals on the land or sell the rights to future development to a conservation organization to prevent the land from someday being turned into a shopping mall.

Living things—plants, animals, microbes—are material property and can be privately owned and traded. A person can buy an ox, use it to plow his fields or rent it out for plowing, mate it and keep or sell its offspring, or slaughter it and eat or sell the meat. Human beings are animals biologically; however, the human body, with its parts and products, is generally accorded a status different from that of other material property. Living human beings are understood to own themselves, although standards differ on exactly what rights this ownership implies. For example, historically, there has been controversy over whether one could sell oneself or buy someone else; today, slavery is almost universally condemned. It is acceptable to rent oneself out for many kinds of work, but it is socially stigmatized and often illegal to sell one's sexual services. Attitudes toward trade in human body parts and products are mixed: selling a renewable body product like hair, blood, or sperm is more likely to be found acceptable than selling a kidney.

The status of human bodies after death raises special issues. The owner is dead, but the body *en toto* does not exactly become material property that is owned by someone else (e.g., the heir to the estate), who can do whatever he or she wants with it. Respectful treatment of dead bodies is an important cultural value. The definition of respectful treatment differs across cultures, varying with religious views, the wishes of the deceased and the family, and other social factors. The persistent strength of this value is, however, evident in such phenomena as Egyptian mummies, the horror aroused by body snatchers in Victorian England, and the relentless search for the remains of American soldiers missing in action and the victims of the 9/11 destruction of the World Trade Center.

It seems clear that the human microbiome can reasonably be considered material property. It is composed of living material things—the living organisms found on or in human bodies. Every person has a microbiome inhabiting his or her body, and the person can have economic property rights in his or her microbiome in the sense of the ability to use, give, or trade samples of it. Thus, a specific person's microbiome can be considered private material property, and society can choose to recognize a legal property right in it. Specifying the exact nature of the legal property right can be complicated, however, when there is no consensus on whether a person's microbiome is an integral part of the body or separate from it. When it is regarded as the body, it is arguably subject to the confusing and somewhat arbitrary legal rules that govern property rights in bodily parts and products. To complicate the issue, scientists doubt that current technology can completely separate a person's human DNA from the DNA of the collection of bacteria, viruses, and fungi that composes the person's microbiome. In other words, even if the microbiome is conceptually distinguished from the body, it may not be separable in practice. This could complicate treating microbiome samples as property or impose special restrictions on their use or exchange.

Sometimes people speak of the human microbiome as common property. What this means is not entirely clear. It cannot mean that the human microbiome is common material property. In concept, the human microbiome is a material thing, namely, the aggregate of the microbiomes of all humans. At any point in time, however, each human being has exclusive use of his or her own physical microbiome in the sense that the same microorganism can't be on two humans at once. If there are commonalities in the types of organisms that make up the human microbiome across subgroups of humans, or across all humans—if, for example, there turns out to be a core microbiome that is common to all human beings—this information is better understood as an element in a classification scheme than a commonly owned piece of material property called the human microbiome.

In contrast to the tangible things called *material* property, the term *intellectual* property refers to intangible ideas. Research on the human microbiome will produce a growing body of knowledge about what the human microbiome is, how it functions, what role it plays in human health, how it influences and is influenced by the physical environment, what kinds of organisms compose individual microbiomes and what role they play physiologically, and so on. To the extent that this knowledge is about the nature of biological processes and how the world works, it does not involve exclusive use, and such knowledge has traditionally been seen as *common* intellectual property. This suggests a metaphorical sense in which the human microbiome could be spoken of as common property. We all "own a piece" of the (aggregate) human microbiome. Our piece (of material property) both influences and is influenced by all the other pieces, and the scientific knowledge of the human microbiome and how it functions is the common IP of everyone.

In the research context, the major property rights controversies concern private IP rights that are established and protected by law. This protection is not extended to basic scientific ideas, such as ideas about the role of the microbiome in evolution, or to information that is simply descriptive, such as data on the microbial composition of individual human microbiomes and its variation across groups defined by location, ethnicity, and other variables.

Rather, protected IP is the applied knowledge of how to do or make something, or creative products like a musical composition or a literary work. The three major types of IP protection are patents, copyrights, and trademarks. Patents relate to products and processes, copyrights to musical compositions and written work, and trademarks to the use of specific names and marks to identify the seller of a good or service. Patents and copyrights are the substantive types of IP that are potentially relevant to microbiome research.

Patents

Research on the human microbiome will result in the identification of microorganisms and the development of new products and processes that may be eligible for patents. To see the difference between private ownership of *material* property and private ownership of *intellectual* property in the form of a patent, consider the invention of the wheel. The maker of the first wheel owns the wheel he makes; it is his material property to use or to sell. He can make more wheels, and they remain his until he sells them, gives them away, or abandons them. If the inventor dies, ownership of the remaining wheels passes to his heirs. But suppose another person makes a wheel, by buying one of the inventor's wheels to use as a model, or by simply copying it from observation. The concept of IP introduces the notion that the inventor can own *the idea of making a wheel*, in the specific sense that no one else is allowed to make or sell wheels without the inventor's permission, even if the person comes up with the idea of a wheel independently. The inventor would obviously find it difficult to enforce this property right himself.

To secure a property right for a limited time period and government enforcement, the inventor obtains a *patent*. Under U.S. law a person obtains a patent by describing the invention in detail; demonstrating that it is useful, novel, and not obvious to an expert in the relevant field; and showing that he has actually built a working version of the invention. The basic subject of a patent is the idea of a new thing, like the wheel. One can also patent a process: for example, *the idea of a new and nonobvious way to make a wheel*. The patent gives the patent holder a time-limited monopoly on his idea.

Should *living* things be patentable? The standard argument for a social policy that recognizes property rights in intellectual property is that society as a whole benefits from new ideas. IP protection enables those who come up with new ideas to profit from them, and the prospect of this profit is intended to encourage people to produce more new ideas. If IP is about recognizing private property rights in new ideas, this would seem to rule out the patenting of anything, living or nonliving, that already exists in nature. Admittedly, human beings do sometimes modify natural things in novel, nonobvious ways. Over time, patent systems have been expanded to incorporate some instances of this, for example, to allow patenting of hybrid plant varieties and genetically modified bacteria.

Recently, patent law has gone further and taken the more controversial step of expanding its scope to allow patents on some things that have not been modified but only identified: for example, individual human genes. This sets a potential precedent for microbiome researchers who want to patent organisms or combinations of organisms identified in their research.

The goals of IP protection sound good: reward inventors, get more inventions. But there is a downside. A patent creates a government-sanctioned monopoly. This enables the patent holder to charge a higher price for the patented item than would prevail in a market in which competition was allowed. In fact, a strong case can be made that intellectual *property* rights are more accurately called intellectual *monopoly* rights (Boldrin and Levine 2008). There is an inherent tension between the goal of encouraging people to be inventors and the goal of ensuring that the inventions, once made, are broadly available at affordable cost.

PATENTING PROS AND CONS

In the United States, patents are issued by the US Patent and Trademark Office (USPTO). A patent gives an inventor the right to exclude others from making, using, offering for sale, or selling the invention in the United States or importing the invention into the United States. The three categories of patents are utility patents, design patents, and plant patents, with effective terms of 20, 14, and 20 years, respectively. Utility patents cover "any new and useful process, machine, article of manufacture, or composition of matter, or any new and useful improvement thereof" (USPTO 2011, 1). Design patents are for "a new, original, and ornamental design for an article of manufacture" (USPTO 2011, 2). A plant patent is available to "anyone who invents or discovers and asexually reproduces any distinct and new variety of plant" (USPTO 2011, 2). Again, the patented item must be useful, novel, and not obvious to an expert in the field. The patent is effective only in the United States and its territories and possessions. To be the exclusive seller in other countries, the inventor must get patents in those countries. This is a complex process, since the basic rules for what is and isn't patentable vary across countries.

The transaction costs associated with establishing and enforcing patent rights are substantial. The patent office must interpret the legislative language to set specific boundaries on what is patentable and then apply them to the applications. The USPTO processes over 450,000 applications a year and has 6,500 employees. The application process is a costly and time-consuming process for the applicant. Detailed information must be submitted and application fees paid. It is possible to apply on one's own, but the patent office recommends that applicants have specialized help from a lawyer or a patent agent.

After a patent has been issued, the legal system is responsible for enforcement. Patent holders incur costs in bringing suits to defend their rights, society incurs costs in adjudicating legal cases, and other potential inventors incur costs to ensure that their work does not infringe on existing patents. All of this can require significant effort, because a single invention may involve many parts and processes, all and any of which may or may not have patent protection. Moreover, even after a patent is issued, there is often substantial uncertainty about its scope and validity, uncertainty that is only resolved through costly litigation (Eisenberg 2011).

The opposing views on patenting can be summarized as follows:

Patenting is good. In fact, it's vital. If we don't give inventors an incentive to innovate, we will get less innovation in the economy. Also, inventors deserve to be rewarded for their good ideas. The patent system increases innovation and rewards invention by giving an inventor a time-limited monopoly on the patented item.

Patenting is bad. It makes things worse, not better. That's because monopoly (even time-limited monopoly) is bad. A monopolist limits market supply to gain monopoly profits. Once something useful has been invented, it should be widely available at the cost of production plus a standard profit on the investment in production facilities. Consumers deserve to be protected from the high prices and inefficient allocation of resources that monopoly brings.

One curious feature of debates about intellectual property rights is worth noting before we discuss the merits of these two views. Those who are most in favor of free market capitalism as the way to efficient production and low prices—who fiercely oppose government intervention in markets for the sake of social benefit—are often the same people who strongly favor IP rights and, therefore, patents. This is odd, since one of the early lessons in introductory economics is that monopoly is both unfair and inefficient. The monopolist gets excess profits compared to what he would get if other producers were free to enter the market, consumers pay more than they would in a competitive market, and there is a deadweight loss related to the inefficient allocation of productive resources. Monopoly is the enemy of free market competition, so the last thing a free market advocate should want is a government-imposed monopoly.

Some patenting advocates deal with this contradiction by pointing out that free and competitive markets can't exist without private property rights. In their view, an IP system simply recognizes and enforces a person's private property right in his ideas. Secure private ownership of *material* property is indeed essential to a free market economy. People must own their tangible property so they can engage in voluntary buying and selling in markets, and a structure must be in place to ensure that contracts and private property rights are enforced. But intellectual property differs fundamentally from material property. With or without a patent, a person owns his idea in the sense that he is free to use it as he likes. He can tell people about it or keep it a secret. He can make wheels himself and sell them. He can hire himself out to help someone else start a wheel business. What the patent gives him is the ability to control what other people do with their own material property and their own ideas. This prevents competition in the market instead of fostering it. The patent holder has the right to stop a person from making copies of a wheel purchased from the inventor. He can also stop a person from making a wheel even if that person has never seen the inventor's wheel (i.e., a person who has the same idea independently cannot use it). In sum, the assertion that IP protection is part of the basic framework for free, competitive markets is false. The standard justification for patents is actually a prime example of an argument that government intervention in markets is needed to make sure we don't miss out on important social benefits.

The issue to be resolved is whether the argument for government intervention is sound. The argument is at least plausible. It can take substantial effort and financial resources to come up with a new idea, create a working prototype, and develop it into a product that can be commercially successful. If other people can quickly and easily copy what an inventor has done, the inventor may conclude that it won't be possible to make enough to recoup the development expense before competition brings the price down to production cost plus a standard profit. An invention that would produce long-term social benefits doesn't happen because there isn't the incentive to make the initial investment. This is a form of economic inefficiency. The government gives an incentive to innovate by allowing monopoly profits for a period of time to the first person in.

But when the government awards patents, it trades one kind of inefficiency for another. There may be more innovation, but there is also the inefficiency related to the monopoly (including the substantial transaction costs associated with the patent system). Only an empirical analysis can determine whether IP protection in the form of patenting is justified. Does the patent system actually produce additional innovation, and if so, how much? Given the inefficiency and transaction costs associated with the patent system, is patenting the most efficient way to achieve the desired increase in innovation?

In fact, the case for the social value of a patent system is not nearly as clear as the usual rhetoric suggests.[2] Innovation can and does occur in competitive markets in the absence of patenting. Since entry doesn't occur instantly, it is theoretically possible for the first person to bring a new product to market to have an advantage even in the absence of a patent. In practice, economic history shows that much innovation has taken place in the absence of patents and that the first company in can have a good run before competition drives down the price. For example, the first wheeled suitcase with a retractable handle was sold by a company called Travelpro. This was a simple idea, easy to imitate, and it was imitated by other firms to the extent that wheeled suitcases now dominate the market. Nevertheless, the original company has prospered (Boldrin and Levine 2008, 162). Moreover, many ideas are not that easy to imitate and do not spread that quickly, even in the absence of patents. The inventor knows more about the product or process and can earn an above-standard profit while others invest time and effort in learning what the inventor already knows.

In some cases, a patent system can actually have a negative effect on innovation. As noted, the process of getting a patent takes time and money, and inventors have to be careful not to infringe on existing patents. Patents can make it more difficult to engage in incremental innovation (i.e., the kind of innovation that builds on an initial invention). The monopoly profit is a payment for past innovation, not an incentive for future innovation, and can even act as a disincentive for further innovation. In their discussion of patents, Boldrin and Levine (2008) present numerous examples of inventors who, once they had received a patent, shifted their energy from developing new ideas to policing their monopoly on the old one.

---

[2] For a comprehensive treatment of intellectual property protection in theory and in practice, an excellent source is *Against Intellectual Monopoly* (Boldrin and Levine 2008). It presents the theoretical arguments for and against IP protection, discusses the historical development of IP systems in various countries, and surveys the empirical literature on the impact of IP protection on innovation.

Patenting is a blunt policy instrument in that the size of the incentive (monopoly profits for 20 years) is not calibrated to the amount that is actually required to be an effective incentive to innovate in each specific case. Sometimes an innovative idea builds on a foundation of past work and a number of people are working hard along the same lines, and then someone wins the race to be the first with a patentable prototype. In other cases, the innovative idea occurs as an unexpected by-product of some other activity with no special effort. Some inventions are made by people who seem compelled to invent things without regard to future profits.

Viewed as a reward, the benefit of the patent is arbitrary. Even when there is agreement that there should be a reward for being the first in, or for having a brief flash of insight that has long-term social value, there is no obvious reason why it should be a winner-take-all monopoly for an arbitrary time period.

Drug patenting dramatically demonstrates the efficiency and equity issues at stake in IP protection. The cost of developing and testing new drugs is high, and there is no guarantee that any individual drug will make it all the way onto the market. It is claimed that without patenting as an incentive, we would miss out on hugely beneficial new drugs. At the same time, with patenting, the poor and the uninsured in the United States find it difficult to afford the high prices that are charged for drugs that have patent protection. For most of the developing world population, patented drugs are hopelessly out of reach. Often the production cost is quite low, so once a drug is shown to be safe and effective, a firm could produce and sell the drug at much lower prices and still make a profit. The patent holders themselves could increase their profits by selling their drugs at heavily discounted prices in poor countries; however, they are reluctant to do so given the difficulty of separating the U.S. market from developing world markets. Developing countries could simply ignore the patents and encourage local production. To forestall this, the United States and other industrialized countries have been pressuring developing countries to increase their protection of IP.

Given the negative effects of patenting, it's important to know whether it is really essential to the development of important new drugs. The evidence says that it probably is not. The economic history of the modern pharmaceutical industry shows that it developed faster in countries where patents were fewer and weaker (Boldrin and Levine 2008, 215). Surprisingly, a substantial fraction of the drugs that make the list of especially important drugs do not owe their existence to patenting (Boldrin and Levine 2008, 230). At the same time, patenting encourages the development of "me too" drugs that are just different enough from existing drugs to qualify for new patents. It is questionable whether such drugs have enough social benefit to justify granting them patents (Boldrin and Levine 2008, 231–232). Patent monopolies are time limited, but pharmaceutical companies have been very aggressive in using their economic power and the legal system to postpone entry of generic competitors into the market when a profitable patent expires (Kanter 2009; Saul 2008). The pharmaceutical industry has consistently been one of the most profitable industries in the United States, despite its large expenditures on research and development (Boldrin and Levine 2008, 226). Overall, the effect of patenting on the production of truly innovative new drugs is smaller than the rhetoric suggests, and it is achieved at a very large cost (Boldrin and Levine 2008, 238).

There are other ways to encourage innovation besides the patent system. For example, government and private philanthropists can subsidize research and development activity directly and then make the information freely available to any who want to produce and sell the products or use the processes. They can also offer fixed-amount prizes for socially beneficial inventions.

Viewed from the perspective of the American taxpayer, the heavy reliance on patents to encourage innovation can seem perverse. The government funds basic biomedical research and encourages the funded researchers to apply for monopolies on the processes and products related to the research it has funded. Then, as the major funder of this research, it pays monopoly prices for the patented items that are used in subsequent research. Also, as a major funder of personal health care through the Medicare, Medicaid, and Veteran's health programs, the government pays monopoly prices for patented health care items such as prescription drugs, genetic tests, and medical devices. Innovation may be encouraged, but in the process, substantial government revenue leaks out to support comfortable lifestyles for those who have gained title to the monopoly profits. The people who do basic and applied research and development deserve to be paid for their work, but it is not obvious that the best way to do this is to award them monopoly rights in the products and processes that result.

EXPANSION OF PATENT PROTECTION

In recent years, there has been expansion of the patent system in the categories of items covered, the duration of the protection, and the geographic area covered. One type of expansion of particular relevance to this chapter is the aforementioned extension of patenting to things that occur in nature, such as plants, microbes, and genes. Originally, the U.S. patent system did not allow the patenting of living things. This changed with the decision to allow patenting of some new plant varieties. An important further step was taken in 1980 when a court decided, in the case *Diamond v. Chakrabarty* cited at the start of this chapter, that a microbe could be patented (Robinson and Medlock 2005). The microbe had been genetically engineered to break down crude oil and thus could be used to clean up oil spills. To the extent that a plant or a microbe is the product of human manipulation, this makes some conceptual sense, because it can be considered to be a manufactured product.

Expanding patenting protections to allow naturally existing genes and microbes to be patented seems harder to justify. Gene patenting has been very controversial. There is a substantial literature on what it actually means to patent a gene and the conceptual and practical arguments for and against it (see, for example, Eisenberg 2002, and commentaries by Sagoff, Nelkin, Wilson, Quigley, Rai, White, Woessner, and Robertson published with it; also Carbone et al. 2010; Colaianni et al. 2010; Cook-Deegan 2008; Heaney et al. 2009; Holman 2012).

In 2009, a group of medical societies, doctors, and patients brought a lawsuit claiming that genes should not be patentable because they are natural entities (*Association for Molecular Pathology et al. v. U.S. Patent and Trademark Office, Myriad Genetics, et al.*). Lawyers from the American Civil Liberties Union served as counsel for the plaintiffs. The gene patents of Myriad Genetics were chosen for the lawsuit because Myriad Genetics seemed to be a particularly egregious case of using monopoly power in ways that were not in the public interest (Heaney et al. 2009). As summarized at the beginning of this chapter, the company had patents on two genes related to breast and ovarian cancer (BRCA1 and BRCA2), and it had developed a test for the presence of the genes. The company refused to license its diagnostic test to health care providers, so providers had to send patient samples directly to the company for testing. For this service, the company charges a very substantial fee. It also used its patents to prevent alternative tests from being developed. The court decided in favor of the plaintiffs, but the decision was reversed by the Court of Appeals for the Federal Circuit (CAFC). The plaintiffs appealed to the Supreme Court, which vacated the CAFC ruling and sent the case back to the appeals court for further consideration in light of a new high court ruling in a related case, *Mayo Collaborative Services v. Prometheus Labs, Inc.* (Pollack 2012).[3] The CAFC ruled in favor of gene patenting again, and the plaintiffs appealed the case back to

---

[3] Quotations from the Supreme Court's opinion in *Mayo Collaborative Services v. Prometheus Laboratories, Inc.* 560 U.S. ___ (2012) (U.S. Supreme Court 2012): "[L]aws of nature, natural phenomena, and abstract ideas" are not patentable. *Diamond* v. *Diehr*, 450 U.S. 175, 185 (1981); see also *Bilski v. Kappos*, 561 U. S. ___, ___ (2010) (slip op., at 5); *Diamond v. Chakrabarty*, 447 U.S. 303, 309 (1980); *Le Roy v. Tatham*, 14 How. 156, 175 (1853); *O'Reilly v. Morse*, 15 How. 62, 112–120 (1854); cf. *Neilson v. Harford*, Webster's Patent Cases 295, 371 (1841) (English case discussing same). Thus, the Court has written that "a new mineral discovered in the earth or a new plant found in the wild is not patentable subject matter. Likewise, Einstein could not patent his celebrated law that $E=mc^2$; nor could Newton have patented the law of gravity. Such discoveries are 'manifestations of... nature, free to all men and reserved exclusively to none.'" *Chakrabarty, supra*, at 309 (quoting *Funk Brothers Seed Co. v. Kalo Inoculant Co.*, 333 U.S. 127, 130 (1948)). "Phenomena of nature, though just discovered, mental processes, and abstract intellectual concepts are not patentable, as they are the basic tools of scientific and technological work." *Gottschalk v. Benson*, 409 U.S. 63, 67 (1972). "And monopolization of those tools through the grant of a patent might tend to impede innovation more than it would tend to promote it...," pages 1–2.

Still, as the Court has also made clear, to transform an unpatentable law of nature into a patent-eligible *application* of such a law, one must do more than simply state the law of nature while adding the words "apply it", page 3.

Available at http://www.supremecourt.gov/opinions/11pdf/10-1150.pdf. Accessed April 25, 2012.

the Supreme Court, arguing that the CAFC did not apply the Supreme Court's ruling correctly. As of November 2012, the final outcome of the case remains undetermined (Saunders 2012).

In discussions about whether a new item should be patentable, the conversation is often limited to comparisons of the item with other things that are currently patentable. It is implicitly assumed that patenting is basically a good idea. But this is far from clear. As we have explained, a strong case can be made that patenting is not an effective, efficient, or fair government policy even for the things it has traditionally covered, let alone controversial things like genes.

Identification of genes and exploration of their functions produces basic natural science knowledge that can lead to practical applications. This is a good example of the kind of basic research that was traditionally funded directly by the government, private institutions, and philanthropy and made freely available to the scientific community. Moreover, the patent system already provides an incentive to study genes and their functions in the hope that a patentable product will eventually result, even if the genes themselves cannot be patented. Conceptually and practically, it may be better policy to prohibit gene patents and make knowledge about genes and gene functions freely available for use.

A definitive Supreme Court ruling in favor of gene patenting will lead to pressure to extend patenting to the new biological entities that are identified in the context of microbiome research. It may be wise policy to resist this pressure and craft better policies to serve the goals that patents are supposed to serve.

Copyrights

SCIENTIFIC IDEAS

Microbiome research will produce new knowledge, and some of it will be published in copyrighted articles and books. A copyright, however, gives its owner exclusive rights only to the specific form in which ideas are expressed, not to the ideas themselves. Many scientific articles are published in journals and are copyrighted, but the authors generally do not hold the copyrights. Typically, the author signs a release assigning the copyright to the journal as a condition of publication. As a result, the author has restrictions on producing and distributing additional copies of the article but is free to continue to use the ideas it contains. In fact, anyone who wants to can paraphrase the ideas in the copyrighted article, explain them in different words to others, and use them in his own projects. Copyrights do impede the free flow of information to an extent, given the cost in time and money that people must pay to subscribe to journals, consult them in libraries, or obtain legal electronic or paper copies. Electronic distribution through the Internet has, however, dramatically reduced the marginal cost of disseminating written documents.

The importance of copyrights in science is less than it is in other fields. The federal government plays a major role in scientific research, and documents that are directly produced by the federal government are automatically placed into the public domain (they cannot be copyrighted by a private entity). Increasingly, government documents are available free online, and this includes publications on scientific work done within the government. Publications based on research done with federal funding but not by the government itself can be copyrighted by individuals or their organizations (depending on the organization's terms of employment), but as the funder, the government can exert pressure to make results broadly available. There are also private efforts to put more scientific articles in the public domain soon after they are written and to find other ways to allow new scientific information to spread quickly.

SCIENTIFIC DATA

Sharing scientific data is a central element in international human microbiome research collaboration. In the statement of its goals, the IHMC says, "The Consortium's efforts are focused on generating a shared comprehensive data resource that will enable investigators to characterize the relationship between the composition of the human microbiome (or of parts of the human microbiome) and human health and disease."[4]

The scientific data that are developed in the course of research and constitute an essential input into the scientific process are not eligible for copyright. This does not mean that all data are readily accessible. When a data set is made public, even in a copyrighted article, it can be used by others, but the original physical embodiment of a data set is material property and the original owner can choose not to disclose the data. Eisenberg (2006, 1023) points out that:

> . . . in the absence of statutory protection such as a patent or copyright that survives beyond disclosure, a standard strategy for preserving the commercial value of data and databases has been secrecy or, more accurately, restricted access. Some valuable databases are used only internally within a firm, protected under the law of trade secrecy, or made available only to paying subscribers under the terms of database access agreements protected under the law of contracts or through technological restrictions on access to a website.

---

[4]IHMC website at http://www.human-microbiome.org/.

In scientific research, there has always been tension between norms of secrecy and sharing. Scientists are expected to publish their data so that their results can be verified, they are encouraged to share data to advance the overall scientific enterprise, and they are aware that sharing data with others makes others more likely to share data with them. At the same time, even when financial profit has not been an issue, scientists have at times restricted access to data sets they have developed to preserve their ability to exploit them in the competition for publications and academic prestige.

Technological advances in information handling have had an impact on the way the conflict between secrecy and sharing plays out. Copying machines have made it much easier to reproduce printed data; laptop computers can hold and manipulate databases containing vast quantities of data; and the Internet makes it possible to access data easily from remote sites. Thus, technology is facilitating easier access to data. Length restrictions imposed by journals used to limit the amount of supporting data disclosed in an article. Now, journals can easily require authors to post the data online.

The federal government plays an important role in fostering data sharing. The government is directly responsible for the creation and management of many large scientific databases as well as databases of information collected for other social purposes (e.g., Census data, weather data, Medicare and Medicaid administrative data, data used to construct national income and employment statistics). In general, it is federal policy to make data available for free or at the marginal cost of distribution (subject to privacy-related restrictions). The federal government also makes data sharing an important element in the nongovernmental research that it funds.

At the same time, there are also forces that oppose sharing data. The increased emphasis on securing IP protection for products and processes has increased the perceived value of databases that might lead to them, thereby creating new incentives to control access to data. The same technology that makes it easier to access electronic data can also be used to restrict access to it. Not surprisingly, there have been efforts to establish databases as IP that is eligible for protection through patents or copyrights. So far, the efforts have been unsuccessful in the United States (Eisenberg 2006; Reichman and Uhlir 2003).

Finding the right mix of public and private policies—the mix that appropriately balances the benefits from data sharing against the benefits from secrecy in a way that is politically feasible—is a challenging task. It will involve a complex mix of government grant policies, legal rules, and social norms.[5]

---

[5] For discussions of various approaches to data sharing, see, for example, Eisenberg (2006), Gitter (2010), Knoppers et al. (2011), and Reichman and Uhlir (2003).

Property Rights and Biological Samples

Although the involvement of biomedical researchers in entrepreneurial, for-profit activity is a factor in the controversy over property rights in biological samples, it's not *all* about money. The special sense of ownership that people have about their own bodies is also relevant. The sense of being wronged has a different quality when one's body is involved than it would have if more ordinary material possessions were at issue.

"People are using parts of *me* to do things I don't approve of, or at least should know about before they are done." This kind of reaction is particularly evident in the concern people have about the use of their genetic material, which many seem to view as "the essence of me." This characterization is a stretch both scientifically and philosophically, but it has psychological and political force. People also have concerns about privacy and confidentiality that would not arise with ordinary material property (see the chapters on privacy and biobanking).

Nevertheless, the perception that people are making money from the samples does add intensity to the debate and cause people to focus on property rights and being owed a share of the profits as a matter of right. Yet, establishing that biological materials are the material property of the people they come from wouldn't necessarily have the effect of ensuring that those people share in any profits from research results.

In the case of the human microbiome, it can be argued that the organisms that make it up are not actually part of the body, but are simply living things that are on or in it. So for the moment, we will stay away from the complexity of the "it's me!" assertion and focus only on the "it's mine!" The microorganisms that make up the microbiome can be considered material property in the same sense that geese can be material property. Assume that society chooses to recognize the microorganisms on or in the body as the legal private material property of the host, who can use or trade them. If a sample of a person's microbiome is taken with consent, the natural default position would be that the host gives up ownership of the microbes. The consent doesn't necessarily have to be *informed* consent, in which the taker tells the source exactly what he or she plans to do with the sample and what risks and benefits that implies for the person providing it. There is no requirement for informed consent when ordinary property changes hands by sale or gift. The individual who provides the microbiome sample can certainly ask questions about what will be done with it and refuse to proceed if the answers are not to her liking. In the absence of an explicit contract, however, the person who provides the sample would usually have no legal recourse when the new owner uses the sample differently than described.

Suppose the sample leads to something profitable: for example, a researcher identifies a new microbe in the sample that turns out to be a useful ingredient in a product or process used in research or the treatment of a disease. The researcher can culture the microbes in the sample and there is no reason to think that the person whose sample it is has a property right in the descendants of the original microbes (just as the owner of geese would have no property right to the offspring of geese he sold or gave away). If the host body turns out to be a uniquely suitable environment for the microbes and the researcher wants more of them, the researcher can try to acquire additional samples. If the original host is aware of their value to the researcher, she can try to extract a higher price. This effort may not be very successful when many people host the same microbe. But there is no reason to think that the researcher owes the original host any ongoing payments based on the fact that it was her sample that allowed the researcher to learn the value of the microbe.

To see this, compare this situation to that of a person who has a wood lot and sells lumber to an inventor who uses it to make the first wheel. Supplying the wood doesn't automatically give the wood lot owner a right to an ongoing share of the profits from selling wheels, even if the maker of the first wheel obtains a patent on it. This is true even if one of the logs sold happens to have special properties that make the wheel possible. Of course, once the wood lot owner realizes that some of her lumber has the special properties, she may be able to get a higher price for those logs. In this way, she will benefit from the invention of the wheel by supplying the special logs to wheel producers. But other wood lot owners may have the special logs, or if they don't, they may start growing them. Where the price eventually settles will depend on the usual factors of supply, demand, and production cost that determine long-run prices in a market economy.

Now suppose the person who invented the wheel manages to patent not only the wheel but also the special logs. The logs exist in nature and the inventor has done nothing to create them, yet he now has exclusive use of them in the sense that no one else can use them without the permission of (and perhaps the payment of royalties to) the patent holder. Other owners of special logs find themselves restricted in the use of their own material property. In the microbiome application, this corresponds to the situation in which the researcher obtains a patent on the microbe itself, and it illustrates why assigning a person the private IP rights to a thing that exists in nature is problematic. Many people have an intuitive feeling that the person who provided the microbe has at least as much right to earn royalties as the person who took it and patented it. One solution would be to pay royalties to the original owner of the microbe. A more intellectually coherent solution would be to declare that things in nature cannot be patented, and neither party should be able to earn royalties on the microbe itself.

Instead of assuming that the microbiome sample is taken from the body by the person who wants to use it, suppose we assume that the sample has been separated from the body and discarded or transferred in such a way that the original owner has no expectation of repossession. For example, suppose a person leaves samples of his microbiome in saliva on a discarded coffee cup, feces in a public toilet, or sweat on a public surface. The usual assumption in property law would be that the person no longer has property rights in the sample and thus no say over what is done with it and no claim on any profits that result from it.

Suppose a microbiome sample is taken for clinical care purposes with consent. As noted earlier, the default position would be that the patient no longer has property rights in the sample. Patients may not realize this, and when they learn that leftover samples are being used for nonclinical purposes, they may not be happy, especially if at the same time they learn that someone else is making money from them. People don't expect to get their leftover blood samples, biopsies, cheek swabs, or fecal samples returned to them as they leave the hospital, but they may expect that the samples will either be stored for future use for their clinical benefit or destroyed.

If parties involved in a transaction find the default policy unappealing, they may try to establish a different arrangement by negotiating a contract. The contract could specify in detail exactly which property rights are being transferred when a piece of material property is traded. For example, a patient could negotiate a contract that specifies how the microbiome sample may or may not be used or specifies how profits earned as a result of using the sample will be shared. Such a contract would incur transaction costs, and it is by no means clear that the patient would find anyone who wanted to participate in such a complicated contingency contract. Alternatively, health care institutions could establish an explicit policy about what would be done with microbiome samples and make consent to the policy part of the patient's consent to treatment. If patients did not like the policy, they could go elsewhere for care.

The government could play a role by establishing and publicizing an explicit default policy on legal ownership and use of leftover microbiome samples. For example, three possible policies are (1) the health care providers own them and can use them for any purpose, (2) the providers own them but can use them only for the patient's clinical benefit or for research, or (3) the providers own them but can use them only for the patient's clinical benefit or destroy them. The government could permit voluntary contracting by patients and providers who wanted terms different from those in the default policy, or it could make the default policy binding on everyone. If the government decides that legislative intervention is justified, it should make rules that balance the interests of individuals and the broader social good. Transaction costs—their size and distribution—are very important in finding the right balance. What the default rules are will affect the transaction costs and make certain kinds of activities too expensive.

We have argued that assigning standard private material property rights to microbiome hosts wouldn't automatically give them complete control over what is done with samples of their microbiomes or shares in any profits that result from the samples. Turning from the "it's mine!" issue to the "it's me!" issue, we can consider the implications for control and profit sharing if the microbiome is considered an integral part of the body. Because of the special sense of ownership that people have about their own bodies, there is precedent for treating human body parts and products differently from standard private material property. The existing rules provide no settled framework that can be applied to biological samples. Nevertheless, since the rules that determine property rights are socially constructed, the government, through legislation or legal decisions, can give people more or less control over the use of biological samples. It can also determine the extent to which people are entitled to share in profits made from the use of their samples. New rules should be coherent and consistent and should balance the interests of individuals and the broader social good.

To illustrate the issues discussed previously, we turn to two real-world examples described at the start of this chapter that have implications for the microbiome as property.

HENRIETTA LACKS AND THE HELA CELLS

The case of Henrietta Lacks has caught the imagination of the public. A common take on the case is this: Henrietta Lacks was poor and black. They took her cells, her property, without her permission and used them to create a cell line that made them millions of dollars. At the same time, Henrietta Lacks's children grew up motherless and poor, unable even to afford health insurance (Margonelli 2010; Roston 2010; Sharpe 2010).

Mrs. Lacks had cervical cancer and was treated at Johns Hopkins University, a major research hospital. In 1951, in the course of a medical procedure, samples of her cancerous cells were taken for research purposes. No one asked Mrs. Lacks for her consent. She was not singled out; at the time, her physician was routinely having samples taken from patients with cervical cancer in the hope that a cell line could be developed that would enable resolution of a controversy about the best treatment for this disease. If the cells were in fact patient property, all of the patients should have been asked for permission to take research samples, but they were not asked.

When Mrs. Lacks's cancer cells were cultured, they proved to have an unusual characteristic: the cells continued to live and divide long after other cell lines would have died out. Mrs. Lacks died within a year after the sample was taken. The cells derived from her sample became the HeLa immortal cell line,[6] which has been very useful in biomedical research (Skloot 2010).

---

[6] "Immortal" is the term that was informally applied by researchers because they seemed to live so much longer than other cells. Actually, we have no idea whether or not they are immortal. Henrietta Lacks was diagnosed with cancer and the cells taken in 1951; so far, the cell line has been in existence for less than a normal human lifespan in the United States.

No one asked Henrietta Lacks for permission to take the sample, but suppose her doctor had asked. Would this have saved her children from economic hardship? Made it possible for them to have health insurance? It's hard to see how. She could have given permission for the use of her sample or refused. If she said no, no one would have gotten anything—not Henrietta, not her children, not her doctors, not society. If she said yes, she and her family still would have gotten nothing unless she also negotiated for compensation. There was no reason to think that her cells were particularly valuable when they were sampled for research. At that point, even the shrewdest negotiator would have been unable to get much compensation for the sample. In theory, Mrs. Lacks could have negotiated a contingent contract for a share in the profits from any product that was developed from the sample. But since no one had any idea that her cells would ever lead to a profitable product, it seems very unlikely that anyone at Johns Hopkins would have been interested in entering into such a contract.

The HeLa cell line was not patented, so there was no state-sanctioned monopoly on the cell line. The researcher who developed the cell line did not earn profits from it. The cell line was commercialized in that an organization managed it and sent out samples of the cells to researchers in exchange for money. It did, however, cost money to manage the cells, since they had to be kept alive and uncontaminated under controlled conditions, and they had to be packaged properly and shipped to buyers. The first organization, a nonprofit part of Tuskegee University, became overwhelmed as the demand for the cells grew. The work was taken over by a for-profit organization that was able, through skilled and innovative management, to make a profit from the production and distribution of HeLa cells (Skloot 2010). In the absence of a patent, it seems unlikely that large monopoly profits were made.

In 1966, the research community learned that the HeLa cells had contaminated other cell lines, and in 1973, researchers at Johns Hopkins contacted Lacks family members and asked them to provide blood samples. The researchers hoped to use family DNA markers to understand the contamination problem better and also to learn more about the HeLa cells. Family members agreed to give blood. Their consent, however, could hardly be called "informed." Skloot's account leaves the reader with the impression that the problem was not so much deliberate deception but a huge gap in understanding between the family and the scientists, and the scientists' failure to appreciate it. They should have done a better job of informing the family. If family members had understood how useful samples of their blood would be, perhaps they could have extracted some financial compensation, although this is far from certain. In any event, it is hard to see how the original taking of Henrietta's cells constituted financial exploitation of Mrs. Lacks or her children. Henrietta Lacks lived a life of hardship and poverty that ended too soon. Her children grew up without their mother and also had lives of economic hardship. Mrs. Lacks and her family were wronged in many ways—by racism and by the political failure to enact adequate social supports, just health care, and effective public education systems. Mrs. Lacks and her children deserved better, but not because she had "immortal" cancer cells. Surely they would have deserved just as much if her cancer cells had died like other people's cells, or if no Johns Hopkins researcher had ever even tried to grow them (Palmer 2010).

JOHN MOORE'S LEUKEMIA CELLS

The HeLa cells were developed in the 1950s. By the late 1970s, when John Moore sought treatment for hairy-cell leukemia, the shift toward patenting research results for profit was under way. His physician, Dr. Golde, recognized early in Mr. Moore's treatment that his cells would be useful for genetic research (Bovenberg 2005; Lawnix 2012). In the course of treatment, Dr. Golde removed Mr. Moore's spleen and other tissues and, without informing him, retained them for research use. Falsely, Dr. Golde told Mr. Moore that he needed follow-up treatment at UCLA Medical Center. Mr. Moore traveled back to the center repeatedly, believing that he was receiving treatment. In fact, the visits were to collect additional blood and tissue samples. In 1984, a patent was awarded in the name of the Regents of the University of California on a cell line based on cells from Mr. Moore's spleen. Through the patent, Dr. Golde received substantial royalties in cash and stock options from licensing the cell line.

As in the case of Henrietta Lacks, John Moore's cells were taken without his knowledge and consent. In the Moore case, however, the physician believed that the cells were potentially valuable. Dr. Golde engaged in deliberate deception to obtain additional samples from Mr. Moore. He patented the cell line and he profited directly from it. These elements were absent in the case of Henrietta Lacks.

When Mr. Moore found out that his physician was profiting from his cells, he sued the physician, the Regents, and the other licensees. He claimed that the blood, tissues, and cell line were his tangible personal property and his ownership rights were violated when they were taken for medical research without his consent. He also claimed that when Dr. Golde failed to disclose his intention to use his patient for research and economic gain, he breached his duty to provide Moore with the information he needed to give informed consent to his treatment.

Ultimately, the case went to the California Supreme Court, which decided that Mr. Moore had no ownership rights in his tissue samples after they were removed from his body. He did have a cause of action for breach of the physician's disclosure obligations. The Supreme Court was somewhat conflicted about its decision on the property rights question, however, and commented that property rights in removed organs and tissues should be dealt with by the legislative branch, not in an individual court case. One of the judges took the fact that California had a law requiring the destruction of solid organs after removal from a patient to indicate that by law, no one had any property rights in organs such as Mr. Moore's spleen.

In 2003, a new case (*Greenberg v. Miami Children's Hospital Research Institute et al.*) arose that was similar in some ways to the Moore case but had a different resolution (Colaianni et al. 2010; Evans 2006). The Greenberg group, composed of parents of children affected by Canavan disease and non-profit organizations involved in research on it, entered into collaboration with Dr. Reuben Matalon and the Miami Children's Hospital (MCH) Research Institute to identify the gene responsible for the disease. The group collected biological samples and medical data, including genetic information, from Canavan families and provided them to Dr. Matalon. He used them to isolate the gene and, with MCH, filed for a patent on it. The patent was assigned to MCH, which implemented a program of restrictive licensing that benefited Matalon and MCH.

The Greenberg group had expected the research results to remain in the public domain so they could be used to benefit the Canavan disease community. The group filed suit, claiming that Dr. Matalon had breached the duty of informed consent in failing to disclose that he intended to file for a patent and profit from it. The group also claimed that it had a property interest in the biological samples and genetic information that had been provided.

Because Dr. Matalon did not treat any of the patients who supplied information, the Florida court rejected the informed consent claim on the grounds that there was no physician–patient relationship. As in the Moore case, the court ruled that the group had no property interest in the materials provided. The court did conclude that the group had a valid claim for *unjust enrichment*. Unjust enrichment occurs when a plaintiff provides a benefit to a defendant, and the defendant voluntarily accepts it and then realizes gains from it that make it inequitable for the defendant to keep the benefit without paying for it. In this case, the benefit provided is the samples and data essential to the research, and the gain is the patent and associated license fees. The court action did not determine how much the plaintiffs were entitled to receive, and the case was eventually settled between the parties. Thus, the case establishes the legal basis for the claim of unjust enrichment but gives no guidance on exactly how to interpret the term *inequitable* and how to award compensation.

Conclusion

Property law is complicated. It has evolved over time through legislation and decisions in adjudication of individual cases in response to decisions made by different individuals responding to different sets of facts. The decisions involve interpretation of legislative language and the selection of the most relevant among previous cases. As the controversies over property rights in the field of genetics illustrate, the law must be applied to new technology and biologic entities that were not considered or imagined when the laws were written. To this complexity, add the fact that as a society, we are conflicted as to whether human body parts, products, and now human microbiomes should be considered property.

For the purpose of this discussion, the legal cases mentioned in this chapter have been simplified and summarized so as to serve as illustrative examples. Closer examination shows that the facts and the legal reasoning are actually more complicated. Judges frequently disagree about how to conceptualize fundamental issues even when they agree about which party should prevail. The *Chakrabarty* case was a five-to-four Supreme Court decision. The *Moore* case was dismissed in state court, reversed on appeal, and finally settled in the California Supreme Court, with three judges concurring on the majority decision, one dissenting, and one dissenting on some points and concurring on others. The first judge decided the *Myriad Genetics* case in favor of the plaintiffs, the appeals court reversed that result in a two-to-one decision, and the Supreme Court vacated that court's ruling and directed the court to reconsider the case in light of the high court's very recent decision in a related case (*Mayo Collaborative Services v. Prometheus Labs, Inc.*). The appeals court ruled again in favor of gene patenting, the plaintiffs appealed again to the Supreme Court, and as of November 2012, the final outcome of the case had not been determined.

All of this divergence of opinions shows that there is no settled framework of ownership and property rights that can be easily applied to biological samples and research on the human microbiome. It is interesting that in calling for legislative action on defining property rights in removed organs and tissues in the *Moore* decision, the California Supreme Court was effectively saying that property rights are socially constructed and it's time to construct clearer rules about human samples. At this point in the advent of microbiome research, this seems like sensible advice. Whatever those rules turn out to be, they will shape property rights in the economic sense. They will define what people can and cannot do with their bodies, the products of their bodies, and the microorganisms that live in and on their bodies.

Policy Recommendations

At this point, it may be useful to step back from the details of legal cases and look at the broader picture. An important feature of that picture is the clash between the profit orientation that has developed in research and the hostility in research ethics to compensation for the provision of a biological sample or more direct participation as a human research subject. The researchers, biotech companies, and pharmaceutical companies are making large private profits and restricting access to the products of the research. Meanwhile, the people who provide the biological samples or participate more directly as human research subjects are supposed to be motivated entirely by a desire to serve the social good.

One way to deal with this clash would be to relax the rules on compensation for research subjects and allow or even encourage the development of more self-interested participation in research. Society could develop new material and intellectual property structures that give people additional private property rights in the samples they provide and even create structures that allow the providers of samples that turn out to be profitable to share directly in the profits. It could encourage extensive private contracting among all parties involved in biomedical research—funders, researchers, research subjects, individuals who provide human samples, biobanks, health care providers—that spell out the legal and financial rights and responsibilities. This would, of course, involve very large transaction costs. It would also redistribute financial rewards in unpredictable ways, so the equity implications would have to be monitored and evaluated. As noted earlier, it is by no means obvious that researchers or donors of biological samples deserve to earn profits on research results. In this model, the science commons would shrink, which might, in turn, have a negative effect on the direction and pace of scientific progress.

Another way to respond to the property issues involved in microbiome research would be to see research as a collective enterprise in which everyone cooperates to produce new knowledge and new processes and products that serve the good of society. In this model, research data would be shared. Policies would be adopted that would reduce the emphasis on securing patents on research results, especially on naturally occurring entities. Research results would be considered common property, and researchers would be rewarded for their contributions in ways other than monopoly profits. Contributions to research by nonresearchers that involved no more than minimal risk and inconvenience—for example, the provision of most biological samples—would be seen as part of one's civic duty and a matter of routine. Serving as a human subject in research would be praised and honored as a contribution to the common good (Hawkins 2009). This model might reduce the financial incentives to researchers and fail to convince people to provide biological samples, which might, in turn, have a negative effect on the direction and rate of scientific progress.

The arrangements that eventually develop in the biomedical research environment are likely to fall somewhere between these two models. Reichman and Uhlir (2003, 399) highlight the contradictory messages that underlie the government's current policies:

> One message reminds scientists of their duties to share and disclose data, in keeping with the traditional norms of science. The other, more recent message urges them to transfer the fruits of their research to the private sector or otherwise exploit the intellectual property protection their research may attract.

Managing these contradictory messages is an ongoing challenge. The NIH Human Microbiome Project and the IHMC are structured as collective enterprise projects. They include commitments to releasing data promptly to shared databases and refraining from seeking IP protection on certain types of basic science results: "pre-competitive, basic data... [f]or example, sequence or expression data from a bacterial metagenomic study" (Interim IHMC Steering Committee 2008).

At the same time, the principles do allow researchers to seek IP protection on other kinds of results: "... data from follow up studies of the functional role of the metagenomic bacterial community or individual bacteria in that community" (Interim IHMC Steering Committee 2008). Also, they include publication guidelines designed to respect the interests of the investigators generating the data.

As microbiome research goes forward, there will be a constant tension between the norms of the science commons and the pursuit of profit through intellectual property protection. We recommend that whenever possible, public and private policymakers:

- Choose policies that support a science commons model, with results and data shared freely and reduced emphasis on securing patents on research results, especially on naturally occurring entities
- Deal with human biological sample consent and compensation issues within a framework of civic participation in a collective enterprise rather than one of intellectual property rights in the samples

Finally, we note that the public reaction to the Henrietta Lacks book provides an important insight. It would be much easier to convince the public that biomedical research is a collective enterprise if we had a health care system that guaranteed everyone access to a fair share of health care no matter what happened to their health or financial position. The fact that health insurance keeps coming up in comments on the Henrietta Lacks case shows that Americans see a definite connection between supporting medical research and having guaranteed access to its results. When the individuals who provide human biological samples are able to count on a fair share of health care and a fair chance of sharing in the benefits of medical advances, they might be more willing to accept a system in which biological samples are routinely used for biomedical research with confidentiality protection but without specific consent or compensation.

References

Barzel, Yoram 1997. *Economic Analysis of Property Rights* (2nd ed.). New York: Cambridge University Press.

Boettiger, Sara, and Alan B. Bennett. 2006. Bayh-Dole: if we knew then what we know now. *Nature Biotechnology* 24(3): 320–323.

Boldrin, Michele, and David K. Levine. 2008. *Against Intellectual Monopoly.* New York: Cambridge University Press.

Bovenberg, Jasper. 2005. Whose tissue is it anyway? *Nature Biotechnology* 23(8): 929–933.

Carbone, Julia, E. Richard Gold, Bhaven Sampat, Subhashini Chandrasekharan, Lori Knowles, Misha Angrist, and Robert Cook-Deegan. 2010. DNA patents and diagnostics: not a pretty picture. *Nature Biotechnology* 28(8): 784–791.

Colaianni, Alessandra, Subhashini Chandrasekharan, and Robert Cook-Deegan. 2010. Impact of gene patents and licensing practices on access to genetic testing and carrier screening for Tay-Sachs and Canavan disease. *Genetics in Medicine* 12(4, Suppl): S5–S14.

Colaianni, Alessandra, and Robert M. Cook-Deegan. 2009. Columbia University's Axel Patents: technology transfer and implications for the Bayh-Dole Act. *Milbank Quarterly* 87(3): 683–715.

Cook-Deegan, Robert. 2007. The science commons in health research: structure, function, and value. *Journal of Technology Transfer* 32: 133–156 (published online: December 7, 2006, doi:10.1007/s10961-006-9016-9).

Cook-Deegan, Robert. 2008. Gene patents. In Mary Crowley (ed.), *From Birth to Death and Bench to Clinic: The Hastings Center Bioethics Briefing Book for Journalists, Policymakers, and Campaigns* (pp. 69–72). Garrison, NY: The Hastings Center.

Cook-Deegan, Robert, and Tom Dedeurwaerdere. 2006. The science commons in life science research: structure, function, and value of access to genetic diversity. *International Social Science Journal* 58(2): 299–318.

Eisenberg, Rebecca S. 2002. How can you patent genes? *American Journal of Bioethics* 2(3): 3–11.

Eisenberg, Rebecca S. 2006. Patents and data-sharing in public science. *Industrial and Corporate Change* 15(6): 1013–1031.

Eisenberg, Rebecca S. 2011. Patent costs and unlicensed use of patented inventions. *University of Chicago Law Review* 78: 53–69.

Eisenberg, Rebecca S., and Richard R. Nelson. 2002. Public vs. proprietary science: a fruitful tension? *Daedalus* 131(2): 89–101.

Evans, Paula C. 2006. Legal affairs: patent rights in biological material. *GEN Genetic Engineering & Biotechnology News* 26(17). http://www.genengnews.com/gen-articles/patent-rights-in-biological-material/1880/. Accessed March 26, 2012.

Gitter, Donna M. 2010. The challenges of achieving open-source sharing of biobank data. *Biotechnology Law Report* 29(6): 623–635.

Hawkins, Naomi, Jantina de Vries, Paula Boddington, Jane Kaye, and Catherine Heeney. 2009. Planning for translational research in genomics. *Genome Medicine* 1(9): 87.1–87.8.

Heaney, Christopher, Julia Carbone, Richard Gold, Tania Bubela, Christopher M. Holman, Alessandra Colaianni, Tracy Lewis, and Robert Cook-Deegan. 2009. The perils of taking property too far. *Stanford Journal of Law, Science, and Policy* 1 (May): 46–64.

Hobbes, Thomas. 1651. *Leviathan* (ed. C. B. Macpherson, 1988). London: Penguin.

Holman, Christopher M. 2012. Debunking the myth that whole-genome sequencing infringes thousands of gene patents. *Nature Biotechnology* 30(3): 240–244.

Interim IHMC Steering Committee. 2008. The International Human Microbiome Consortium: a description of its goals, operating structure and principles. A summary of the IHMC organization. http://www.human-microbiome.org/index.php?id=61. Accessed April 15, 2012.

Kanter, James. 2009. E.U. takes on drug companies for allegedly delaying generics. *New York Times*, July 9, Business Day.

Knoppers, Bartha Maria, Jennifer R. Harris, Anne Marie Tassé, Isabelle Budin-Ljøsne, Jane Kaye, Mylène Deschênes, and Ma´n H. Zawati. 2011. Towards a data sharing Code of Conduct for international genomic research. *Genome Medicine* 3(46): 1–4.

Lawnix. Moore v. Regents of the University of California—case brief summary. http://www.lawnix.com/cases/moore-regents-california.html. Accessed March 26, 2012.

Locke, John. 1689. *Two Treatises of Government* (ed. Ian Shapiro). New Haven, CT: Yale University Press.

Macpherson, C. B. 1985 *The Rise and Fall of Economic Justice and Other Papers*. New York: Oxford University Press

Margonelli, Lisa. 2010. Eternal life. *New York Times*, February 7, Sunday Book Review, p. 20.

Nelkin, Dorothy. 2002. Patenting genes and the public interest. *American Journal of Bioethics* 2(3): 13–15.

Nelson, Richard R. 1959. The simple economics of basic scientific research. *Journal of Political Economy* 67(3): 297–306.

Nelson, Richard R. 2004. The market economy, and the science commons. *Research Policy* 33(3): 455–471.

Palmer, Larry I. 2010. Reparations. *The Hastings Center Report* 40(6): c4.

Peterson, Jane, Susan Garges, and 37 other members of NIH HMP Working Group. 2009. The NIH Human Microbiome Project. *Genome Research* 19: 2317–2323.

Pollack, Andrew. 2012. Justices send back gene case. *New York Times*, March 27, p. B1.

Quigley, Rosemary. 2002. Waiting on science: the stake of present and future patients. *American Journal of Bioethics* 2(3): 15–16.

Rai, Arti K. 2002. Locating gene patents within the patent system. *American Journal of Bioethics* 2(3): 18–19.

Rai, Arti K., and Rebecca S. Eisenberg. 2003. Bayh-Dole reform and the progress of biomedicine: allowing universities to patent the results of government-sponsored research sometimes works against the public interest. *American Scientist* 91(1): 52–59.

Reichman, J. H., and P. F. Uhlir. 2003. A contractually reconstructed research commons for scientific data in a highly protectionist intellectual property environment. *Journal of Law and Contemporary Problems* 66(1&2): 315–462.

Rhodes, R. 2009. Hobbes's fifth law of nature and its implications. *Hobbes Studies* 22: 144–159.

Robertson, John A. 2002. Sequence patents are not the issue. *American Journal of Bioethics* 2(3): 22–23.

Robinson, Douglas, and Nina Medlock. 2005. Diamond v. Chakrabarty: a retrospective on 25 years of biotech patents. *Intellectual Property & Technology Law Journal* 17(10): 12–15.

Roston, Eric. 2010. Book review of *The Immortal Life of Henrietta Lacks* by Rebecca Skloot. *Washington Post*, January 31, p. B01.

Rousseau, Jean-Jacques. (1755) 1992. *Discourse on the Origin of Inequality* (trans. Donald Cress). Indianapolis: Hackett.

Rousseau, Jean Jacques. (1762) 1968. *The Social Contract* (trans. Maurice Cranston). London: Penguin.

Sagoff, Mark. 2002. Intellectual property and products of nature. *American Journal of Bioethics* 2(3): 12–13.

Sampat, Bhaven N. 2006. Patenting and US academic research in the 20th century: the world before and after Bayh-Dole. *Research Policy* 35: 772–789.

Saul, Stephanie. 2008. Settlement delays a generic Lipitor for many months, a boon to Pfizer. *New York Times,* June 19, Business.

Saunders, Ruth. 2012. US Supreme Court asked to reconsider gene patents. *BioNews* 675. http://www.bionews.org.uk/page_186210.asp?hlight=Myriad+Genetics%2C. Accessed November 11, 2012.

Sharpe, Virginia A. 2010. One life, many stories. *The Hastings Center Report* 40(4): 46–47.

Skloot, Rebecca. 2010. *The Immortal Life of Henrietta Lacks.* New York: Random House.

U.S. Patent and Trademark Office. 2011. *General Information Concerning Patents.* Alexandria, VA: U.S. Patent and Trademark Office. http://www.uspto.gov/patents/resources/general_info_concerning_patents.pdf. Accessed March 26, 2012.

U.S. Supreme Court. 2012. Opinion of the Court, MAYO COLLABORATIVE SERVICES *v.* PROMETHEUS LABORATORIES, INC. 560 U.S. ___ (2012), March 20. http://www.supremecourt.gov/opinions/11pdf/10-1150.pdf. Accessed April 25, 2012.

White, Gladys B. 2002. Patenting genes? A finger in the dike of a bricks-and-mortar patent system. *American Journal of Bioethics* 2(3): 20.

Wilson, Jack. 2002. No patents for semantic information. *American Journal of Bioethics* 2(3): 15–16.

Woessner, Warren D. 2002. Let's get physical. *American Journal of Bioethics* 2(3): 21–22.

4

Privacy, Confidentiality, and New Ways of Knowing More

Nada Gligorov, Lily E. Frank, Abraham P. Schwab, Brett Trusko

Introduction

Recently, researchers have collected skin microbial samples from computer keyboards and showed that these samples included a microbial mark that could be used to identify specific individuals. This was true even after a significant amount of time elapsed (Fierer et al. 2010). The Fierer et al. study illustrates that the microbiome of the skin not only is specific to each individual but remains relatively unchanged over time. Such results introduce the possibility of using microbial marks in forensics and criminal proceedings.

Additional research shows that people from distinct geographical regions have different microbiota. Thus, it might become possible to identify nationality or ethnicity by analyzing a sample of someone's microbiome (Nasidze et al. 2009). Furthermore, changes in the microbiome could be indicative of past behavior. Travel to particular regions could change a person's microbiome by introducing new bacteria unique to a particular locale. Use of antibiotics also changes a person's microbiome (Fortenberry et al. 2010). Microbiome researchers have even observed that sexual habits can affect the microbiome of the penis, making it possible to trace a man's sexual habits by obtaining a sample of his genital microbiome (Ravel et al. 2011). Based on this evidence, we can conclude that each individual's microbiome reads as a biography, replete with revealing personal information.

Most recently, a study was published presenting data indicating that human gut microbiota fall into three enterotypes (Arumugam et al. 2011). In the future, such information could be used to type people and possibly produce social consequences that could be both positive and negative. On the one hand, information about gut enterotypes could be used in clinical medicine to identify the most effective treatment for each individual. On the other hand, if some enterotypes are less conducive to good health, information about them could be stigmatizing to individuals with that enterotype.

In this chapter, we explore various philosophical approaches to the concept of privacy to see how these views apply to issues of privacy as they might arise from the Human Microbiome Project (HMP). We review three main approaches to privacy: physical or bodily privacy, privacy as it pertains to a private sphere, and informational privacy. We conclude that the types of information that might arise from research on the human microbiome are best treated as a subset of the category of informational privacy.

Furthermore, we argue that confidentiality rather than privacy is a more applicable concept to research. Given that research requires disclosure of information about research participants, the most pertinent issue for the HMP is how to maintain confidentiality of the information obtained in the course of research. We will also discuss some legal guidelines, including the Health Insurance Portability and Accountability Act (HIPAA), as it applies to research, as well as National Institutes of Health (NIH) guidelines for data sharing.

Philosophical Approaches to Privacy

The confidence underlying our everyday use of the term *privacy* could be taken as a sign that the definition of the concept of privacy is settled. The characterization of privacy, however, remains contentious; some have argued that privacy suffers from an "embarrassment of meanings."[1] A number of those meanings will be discussed in this chapter.

Views on the concept of privacy can be reductionist or coherentist. According to the reductionist view, the right to privacy reduces to other rights including property rights or rights over one's body.[2] Because privacy can be reduced to other rights, reductionists argue that privacy does not add any rights over and above other, sometimes better-defined rights over our property or over our body. On a coherentist view, there is something distinctive about privacy that cannot be explained by appeal to any other rights or interests.[3] These two general approaches are attempts to describe privacy.

A descriptive account of privacy should explain what makes something private as well as provide the conditions for what constitutes a loss of privacy. For example, if privacy is defined as control over information about oneself, then privacy could be lost as more people gain access to a piece of information.

---

[1] Kim Lane Scheppele, Legal SE184-85 (1988), as cited in Solove (2006).

[2] The most well-known proponent of this view is Judith Jarvis Thomson, whose view will be discussed in the section on physical privacy.

[3] Proponents of this view include Scanlon (1975) and Rachels (1975).

In this chapter, we will not defend a particular descriptive view of privacy, and we will not attempt to settle the dispute between coherentists and reductionists. Instead, we will assume Moore's (2003) view that privacy is culturally valuable, and that even if one could ultimately reduce our interests in privacy to our interests in our property or the integrity of our body, it is still useful to talk about privacy. Moore argues that even if privacy might be a species of the genus of property rights—just like intellectual property is a species of property rights—it is useful to speak about the species separately from the genus. Furthermore, retaining a distinct concept of privacy accommodates everyday usage of this term, reflecting its cultural importance. Our discussion in the chapter addresses how privacy applies to research and, most specifically, the HMP. Since we are not attempting to settle on a descriptive view of privacy, we will not formulate a view that can identify certain types of information as private. We will instead accept that individuals vary in the kinds of information they deem private.

PHYSICAL PRIVACY

Physical privacy "refers to freedom from contact with others or exposure of one's body to others" (Thomson 1975, 299). According to Thomson, several arguments illustrate and demarcate infringements on physical privacy including infringements on one's property. She illustrates this view with several examples. The first is that of a man who owns a pornographic picture that he keeps in a safe in his home. When you own something, she says, you have both negative and positive property rights to it. For Thomson, privacy in this case can be subsumed under the heading of property rights. For example, the owner of the pornographic picture has the right to destroy it, which is a positive right, as well as the negative right that allows him to prevent other people from having access to his picture. Thomson claims that one of these negative rights is the right "that others shall not look at it." Just because he cannot not protect the picture from being viewed with an x-ray device doesn't mean that he forfeits the right to keep it from view. Similarly, if he had not been able to safeguard his other possessions from a sophisticated burglar, he does not abrogate the right to his possessions (Thomson 1975, 299–300).

We have a strong set of rights over our property, but we have an even stronger set of rights over our person or bodies; Thomson calls these "the right over the person" (Thomson 1975, 305). The rights that we have over our bodies are analogous to property rights, the rights not to be touched or looked at. A right to physical privacy involves restricting access to one's body as well as intimate areas like the home. An infringement on physical privacy might occur when someone makes an effort to observe a person in places where that person has an expectation of privacy, for example, getting undressed in a department store dressing room or using the bathroom. An invasion of privacy also occurs when someone stalks you and your daily movements with the intention of becoming an unseen part of your life, even though some of your movements do not occur in a private space. In addition, physical assault can be construed as a violation of physical privacy, for example, when one's body is touched in an unwanted manner. Being patted down by the TSA, being groped on the subway, or enduring a rectal exam by a law enforcement officer may involve violations of privacy.

It is possible to waive these rights to one's body parts being touched and inspected or our conversations being overheard. For example, when a couple is having a loud fight in their home with the windows open, and someone walking down the street stops and listens, the couple's right to privacy is not violated. Their right not to be overheard, Thomson explains, was waived by leaving the windows open (Thomson 1975, 305–306).

Research on the human microbiome might raise questions about physical or bodily privacy. Insofar as we have some property rights over our bodies, any research that would involve taking samples from a person's body would require at least agreement from the research participant.[4] Whether persons have property rights over their microbiome is discussed in chapter 3, but to the extent that each individual's microbiome can be deemed her property, the rights associated with property would apply to the microbiome. There are, however, potential infringements on privacy that would result from collecting a person's microbiome that would not be accounted for by bodily privacy or rights over property.

To further explain this, we will make use of one of Thomson's examples. Thomson maintains that one could relinquish one's right to privacy, like the couple having a loud fight before an open window. But she also argues that one does not relinquish one's right to privacy if x-ray machines make it possible to view the contents of one's bedroom cabinet. The fighting couple could, with relatively minimal effort, close the window and prevent curious bystanders from overhearing their fight. And because of that we could take the view that the couple had voluntarily relinquished their right to their privacy. To some extent, humans shed their microbiome involuntarily, and do so outside of the confines of their property. Furthermore, it would require great effort and restriction of movement, like never leaving one's home, to prevent spreading microbial marks on everyday objects such as computer keyboards or subway handrails. Using technology to detect microbial residue on chairs or computers and then identifying the particular individual to whom the residue belongs is akin to Thomson's example with the x-ray machine or the sophisticated burglar. Nowadays, there are a number of ways in which what might be

[4]For more on the difference between agreement and informed consent, see the discussion in chapter 7 on Population Studies.

considered private information can be obtained without violating a person's right to her bodily privacy or without trespassing onto her property; fingerprints and DNA are obvious examples. The novel ability to obtain information about persons from their microbial marks is just an addition to this list.

Consider an example for how the concept of bodily privacy, as Thomson describes, does not cover what could for some count as privacy infringements. The example pertains to the use of microbial samples stored in databanks.[5] Once samples are collected from a person, they might be used in countless studies without ever requiring the researcher to acquire another sample from the person and therefore circumventing any need to infringe on bodily privacy. Still, one could conceive of privacy infringements occurring at any point of the research where the samples are not properly protected or kept confidential. For samples that are de-identified, samples initially stripped of personal identifiers of the research participants,[6] infringements on privacy could occur if the data are identified and linked to a medical record. In such cases there is a potential for privacy infringement, but such infringements could not be characterized as assaults on bodily privacy. The infringements would not be on bodily privacy because the participant would not be recontacted or asked for additional samples, yet connecting samples with a medical record makes the research participant identifiable. Identifiability of particular participants could be considered an infringement on privacy especially if the information obtained from research is then shared for nonresearch purposes, such as employment discrimination.[7] Hence, Thomson's account cannot cover possible infringements of privacy that could arise from research on the human microbiome. We then need an account that covers more than just a person's body or property.

THE PRIVATE ZONE

Scanlon (1975) offers another perspective on privacy. For Scanlon, privacy extends beyond the contours of someone's body, or property, to include what he calls the privacy zone. The privacy zone includes one's body and one's property, as well as things such as work cubicles, plots in community gardens, and "personal space." Scanlon argues that all privacy violations have in common is that they involve a violation of the interest that we have in being "free from certain kinds of intrusions." Privacy norms and laws, according to Scanlon, protect the interests that we have in "not being seen, overheard... in having a zone of privacy in which we can carry out our activities without the necessity

---

[5] For more on this, see chapter 6.

[6] The term *de-identified* is further explained in section on Sharing Data.

[7] For a view on how information collection and information processing can lead to privacy infringement, see Solove (2006).

of being continually alert for possible observers, listeners, etc." Although the zone is not easily delineated using physical boundaries, our social norms define a flexible system of limits on being observed by others, a socially constructed "zone of privacy," in which we do not have to vigilantly guard against intrusion. It is the intrusion into the conventional zone of privacy that matters, not just a person's property rights or rights over his person as argued by Thomson (1975). On Scanlon's view, if one decides to bury a valuable and private object in a plot she has been assigned in the local public garden, she has a claim on privacy with regard to that object (Scanlon 1975, 316–318). This means that anybody who decides to dig out the buried object will infringe upon the person's privacy.

Applying Scanlon's view to research on the human microbiome, privacy interests as they might apply in the case of microbial samples would not derive only from the privacy interests we have over our bodies and property, but would extend further to the privacy zone. Moreover, the protection of the zone of privacy, based on Scanlon's view, is not related to ownership. Therefore, even if it is agreed that a person does not have property rights over a microbial sample, she might still have privacy rights insofar as the sample is taken from the person's zone of privacy.

Although Scanlon's view expands the boundaries over which we have claims of privacy, it still cannot account for the kinds of privacy infringements that could arise from the HMP. For example, imagine Jane has been sitting on a park bench for a while, and after she leaves a scientist comes to collect her microbiome from that bench.[8] To obtain a sample of Jane's microbiome, the researcher would not need to invade her body. The researcher would not be taking Jane's property, and thus he would not need her permission. Furthermore, it would be difficult to see how a public park, and any bench in it, could be included in Jane's private sphere, in Scanlon's sense of belonging to Jane's zone of privacy. If Jane spent some time reading on the public bench and decided to leave the copy of her newspaper behind, the next person sitting on that bench would not do anything wrong by taking that newspaper and doing with it as he pleases. This conclusion is supported by views that give us rights over particular objects or tokens, to use Moore's (2003) terminology. If privacy protects our right to particular tokens or objects, purposefully discarding that object then entails that we relinquished our rights to it. The rights or interests associated with microbial marks, however, are not just tied to a particular sample; they also apply to the types of information that one might obtain using the sample left behind. According to Moore (2003), privacy applies to both tokens and types, which means that it includes privacy as it pertains to one's body or one's house, which are objects or tokens, as well as types information about

[8] By "her microbiome" I merely mean to designate the microbiome she left behind on that bench.

oneself no matter how or where such information was obtained (Moore 2003, 218). Discarding an object might relinquish the person's right to that object, but it does not relinquish the interests associated with the types of information that can be obtained about an individual from that object.

Suppose the object Jane left behind was her social security card. In such a case, one would be most likely to say that anybody who notices her social security card would not be justified in making full use of it, for example, using the social security number to uncover information about Jane. Hence, discarding an object does not, in every case, entail that the original owner relinquishes the rights or interests associated with that object. Jane's privacy interests can no longer be associated strictly with the possession of an object; rather, they are also associated with the kinds of information that can be obtained about Jane through the use of her social security card. Furthermore, the infringement on Jane's privacy in this case would not be derived from the transgression into her zone of privacy.

INFORMATIONAL PRIVACY

As the study by Fierer et al. (2010) illustrates, one can obtain a microbial sample from objects a person touched, which obviates the need to collect the sample from that person's body. Hence, if privacy infringements result from collection of microbial marks, they would not be accounted for by Thomson's conception of physical privacy because information about a person through her microbiome can be obtained without infringing on bodily privacy. Furthermore, privacy as it pertains to private property cannot account for the kinds of privacy infringements that could arise as a result of collecting microbial samples. We shed our microbiome not just in our home or our office, but in public places also. For similar reasons, the boundaries of Scanlon's private zone would not cover cases where microbial marks are found in places that cannot be characterized as belonging to the private zone. Thus, in principle, infringements on privacy are possible even in cases where one's microbiome is collected from public places such as park benches. The only account that covers the kinds of potential infringements on privacy that could result from some research on the human microbiome is informational privacy. This kind of view of privacy focuses on privacy of information about a person and leaves open the kinds of ways and places that such information can be obtained.

Fried argues that informational privacy has intrinsic value and cannot be reduced to other rights or interests. Fried defines privacy as "control we have over information about ourselves." Fried argues that this type of control is essential to achieving certain human ends (Fried 1968, 109). Fried argues that privacy is tied to our "integrity as persons" and that without privacy the ability to engage in "respect, love, friendship and trust" is lost (Fried 1968, 477). In this sense, Fried's view of privacy is tied to the concept of autonomy because it emphasizes the importance of persons being able to control information they deem private. This view obviates the need to provide a firm criterion to distinguish between information that is private and information that is not private. Given that ideas about what is considered private are different across cultures and have changed historically, it seems futile to attempt such a general criterion. Emphasizing autonomy and personal control over information and leaving it up to each individual to decide which kind of information is private seems like an amiable approach in a lot of cases.

Our microbial marks may prove to be revelatory of health and lifestyle. Thus, sampling someone's microbiome could be a way of obtaining personal information and preventing each individual from purposefully weaving a story about her life, character, and values and preferences. And if, according to Fried, privacy entails that ability to control information about oneself, then certain kinds of research on the human microbiome could, to some extent, decrease people's ability to control information they consider private. Loss of control could happen just by extending the number of people who are privy to information about an individual.

It could be argued, however, that information obtained from microbial samples contributed to biomedical research need not involve an erosion of privacy relevant to the maintenance of friendship and private relationships, nor will it disrupt our control over our personal narratives. Perhaps the facts obtained from microbial marks will not reveal information that could be harmful in that way.

But Rachels (1975) argues that privacy is valuable to us beyond those situations where the desire for privacy seems to be most acute: in situations where we have something to hide, or situations in which information about us that leaks out against our wishes is potentially embarrassing or damaging to our interests. According to Rachels, to understand the value of privacy, we also need to consider ordinary situations in which the information, activity, or sphere that an individual wishes to remain private is not "embarrassing or shameful or unpopular" and no harm can come to him by having it publically disclosed or revealed. Because we still value privacy even in situations where disclosure would not be harmful, Rachels argues that privacy and secrecy are distinct. Even when people do not have anything to hide, or when information about them will not give cause for others to mistreat them, they may still have an interest in privacy (Rachels 1975, 324–325).

In criticism of Rachels, Francis argues that the exposure of certain kinds of information, or a total loss of privacy about certain kinds of information, may have little if any effect on our ability to maintain a variety of different relationships (Francis 2008, 56). For example, one may wish that others not have access to a zip code and phone number for privacy reasons. Similarly, one's body temperature is a piece of information that a physician must keep confidential, but it is unlikely to have any impact on maintaining intimacy with others (Francis 2008, 56).

One can envision cases, such as the one Francis notes, where information about a person might seem neutral but might interfere with her ability to maintain intimacy with others. For example, if a fact about a person's body temperature is a symptom of sexually transmitted infectious disease, the disclosure of that person's body temperature could compromise her intimate relationships. Breaches of privacy, however, are sometimes justified even if they might interfere with intimate relationships. Thwarting the spread of diseases might suffice as moral justification.

The accumulation of facts about human health might precede our interpretation of those facts. Facts that seem innocuous now could pose a threat to privacy later. The facts about the human microbiome that are currently being collected will be interpreted over time and contextualized in such a way that they might produce facts about human health that could pose risks to privacy. Thus, Francis's retort to Rachels does not successfully establish that some facts are entirely innocuous and cannot threaten privacy.

It is conceivable that the identification and categorization of human microbiota could lead to stigmatization of individuals with unhealthy microbiota or stigmatization of groups with undesirable microbiota. Providing adequate confidentiality protection should prevent such stigmatization; we will discuss some of those protections in later sections. If my microbiome could be used to track my whereabouts, my sexual habits, my travel history, and so forth, others could use that information to judge me and to sometimes stigmatize me. But it is a mistake to confuse potential access to personal information with an actual infringement on privacy. An infringement on privacy only arises when the potential to collect information about me is actualized and misused in some way.

Privacy concerns raised by the HMP pertain to informational privacy. And since different people have distinct ideas about the kinds of facts they consider private, respecting people's informational privacy is a variant of respecting autonomy. As Rachels argues, facts that some might consider private need not be embarrassing or stigmatizing. Thus, judgments about which facts are private should in many cases depend on what each person considers private. In public matters and social policy, lines have to be drawn for everyone, and privacy protections that take into account idiosyncratic views of individual privacy cannot always be offered. Privacy protections in research should focus on the most commonly endorsed worries about personal privacy, such as various kinds of discrimination or criminal prosecution.

Many scientific and technological developments have led to increased access to personal information; the Internet is an obvious example, as it thrives on the exchange of personal information. Infringements on privacy, however, do not follow directly from access; they follow from the unwarranted use of that access. To argue that potential or actual access to private information is not justified, one would have to show that its uses do not follow agreed-upon privacy protection and moral guidelines.

Confidentiality

We have maintained that access to information is not the same as unwarranted use of information. But our increased ability to access information certainly proliferates the opportunities for unjustified infringement. It also contributes, as have other technological and scientific advances, to the reconceptualization of the domain of privacy. Before the advent of x-rays, DNA analysis, fingerprinting, and now human microbial marks, the boundaries of privacy could follow the boundaries of our bodies and property. Furthermore, the advent of the Internet and the proliferation of its use in almost every area of life have made it increasingly difficult to specify the domain of expected privacy. Disclosure of information about oneself has now become a requirement of daily life. There are very few facts about any person's life that remain undocumented.

Most transactions, such as application for credit cards or purchases on the Internet, require disclosure of information about the applicant or the customer. Even the most mundane daily activity such as using a cell phone, sending an e-mail, or using a credit card provides potential access to information about an individual.[9] Most of us assume these risks willingly. If we were particularly concerned about privacy, the cost of preempting potential disclosure of private information would be high. It would require abstaining from many daily activities. It is no longer enough, as in Thomson's example of the fighting couple, just to shut the window.

Because of these rapid developments in science and technology, it is important to adjust our ideas about how we can preserve some of the interests or rights designated by privacy. It is no longer possible for a single individual to safeguard his own privacy. If it is indeed the case that privacy will remain culturally valued, the responsibility for safeguarding privacy will require social collaboration.[10] In some contexts, such social collaboration is captured by the concept of confidentiality. Confidentiality allows for a person to disclose information in certain contexts, but the information is protected, and its use limited, by the obligation to maintain confidentiality. Physicians are known to have that obligation. Furthermore, tort law support the requirement of

---

[9] For a recent update on cell phone surveillance, see "The End of Privacy?" *New York Times*, July 15, 2012.

[10] Solove (2006) presents the view that privacy is constitutive of civil society (p. 488).

confidentiality for any profession that depends on trust (Solove 2006, 524). It seems fitting that confidentiality obligations should be extended to researchers as well because the relationship between research participants and researchers also requires trust.

In medicine, the protection of confidential information is considered essential because patients would be reluctant to disclose private information necessary for medical care if confidentiality was not guaranteed. Several studies document that people are more hesitant to seek care when they fear that their health care information will not be kept confidential.[11] The obligation to maintain confidentiality protects patients in the following ways. First, it prevents information about the patient from being disclosed in ways that could be harmful to the patient; second, it provides health care professionals with valuable information for diagnosing and treating patients. Confidentiality also addresses respect for autonomy because it restricts information sharing to a certain group of people and for a specific use, patient care. Presumably, patients agree to reveal intimate details about themselves only for the purpose of enabling physicians and other health professionals to provide needed treatment.[12] Although the interaction between a patient and a physician may entail a great deal of disclosure, which in turns leads to a loss of privacy, the need for that loss of privacy is justified by the goal of maintaining health. The possible harms of disclosure in the clinical situation are mitigated by the physician's obligation to maintain confidentiality. In addition, confidential patient information should be shared only with those individuals who need to know it in the course of their collaboration in the patient's treatment and who are also bound by confidentiality. Our assumption is then that confidentiality might be important to research participants in similar ways and that protecting confidentiality might encourage research participation.

In clinical care, clinicians aim to directly produce benefits for patients. The justification for disclosing private information to a physician is obvious: the disclosure might result in direct benefit to the patient. Research is often not designed to produce individual patient benefit. Consequently, researchers are more likely to be disqualified from access to confidential information because potential breaches of confidentiality constitute a risk that is not outweighed by direct benefit to the individual participant. Perhaps a better way of ethically evaluating research where potential breaches of confidentiality are considered the only risk of research participation is to weigh those risks against potential societal benefits, not direct individual benefit. Construed in this way, the access to private patient information for research might be justified in relation to the potential benefit to society.

---

[11] These studies include Ford et al. (1997); Cheng et al. (1993); and Asch, Leake, and Gelberg (1994).

[12] Parts of this section are based on the commentary in Schwab, Frank, and Gligorov (2011).

To abate concerns regarding potential breaches of confidentiality, we should aim to minimize them with effective confidentiality protections. We also need to emphasize to researchers that they have confidentiality obligations to protect participant information in a similar way as physicians. This would obviate the need to limit researchers' access to confidential information. By having adequate confidentiality protections, established by law or through institutional policy, we can encourage people to participate in research. Publicizing the confidentiality protections that are incorporated into research practice would engender trust in research institutions.

Legislation to Protect Medical and Research Information

We have outlined how research on the human microbiome might infringe on informational privacy and explained how legal rules and institutional policies could be established to maintain confidentiality. Now we shall review the current legal guidelines that protect information obtained during clinical and research encounters. Although these existing legal protections do not provide complete protection of confidential information about research participants, they do provide significant safeguards. We will point to the ways in which current laws protect confidential information and list some ways in which they fail to protect.

Most of the current rules and policies are described in terms of privacy rather than confidentiality, which we consider to be the more appropriate term. None of them aim to prevent the sharing of information entirely; rather, they aim to narrow the circle of individuals and institutions that have access to the confidential information. One such piece of legislation is the Health Insurance Portability and Accountability Act (HIPAA), which regulates the use of medical information and the use medical information in research (Annas 2003, 1489). As per HIPAA, the privacy of medical information entails that no one, other than the patient, may access private medical information without the explicit consent of the patient.

HIPAA regulations provide some further details for the confidential sharing of private information. Under HIPAA regulations a physician is allowed to view the patient's medical records "if he or she conducts any medical business, including billing, electronically, even if the physician contracts with another entity or business associate to do billing" (Annas 2003, 1487). The guideline for sharing medical information is to share "the minimum necessary" to accomplish the purpose. In addition to federal regulations, HIPAA is supplemented by various state regulations, which govern disclosure of information with regard to minors, mental health, or genetic records.

Under HIPAA, explicit consent is not required for the disclosure of medical records if they are used for quality assessment, performance evaluation, the conduct of training programs, the rating of premiums, auditing, business planning, and management. Instead of consent, the health care provider is required to provide patients with a privacy notice. A written privacy notice must be made available to all patients on the first day of the delivery of health care services and it must be prominently posted at the service site. The privacy notice should include information about who will be able to view the information, the specific uses that will require the patient's consent for disclosure, and what uses will not require consent.

Although HIPAA regulations primarily cover the privacy of medical information, they also provide some guidelines for the protection and disclosure of medical records for research. Medical information can be used for research when the records are de-identified or part of a "limited data set," and if the institutional review board (IRB) permits a waiver of consent. Under Title 45, Public Welfare, Department of Health and Human Services, Part 46, Protection of Human Subjects, or "The Common Rule," an IRB can waive the consent requirement when the only record linking the participant to the research study is the consent form and when that form presents the only risk to privacy.

Another piece of legislation that focuses on issues of privacy and disclosure of medical information is the Genetic Information Nondiscrimination Act (GINA) of 2008. GINA is meant to prevent genetic discrimination for medical insurance and employment discrimination. Under GINA, health plans cannot require genetic testing for coverage and cannot refuse coverage based on genetic information. Furthermore, health insurance premiums may not be increased based on genetic information about the patient.

GINA amends HIPAA in the following ways. It requires that genetic information be treated as medical information. It also prohibits the use or disclosure by group health plan, health insurance coverage, or Medicare supplemental policy of genetic information about an individual for underwriting purposes.

GINA also provides protection from discrimination based on genetic information. The act prohibits an employer, an employment agency, a labor organization, or a joint labor–management committee to fail to hire, fire, or discriminate in terms of employee compensation based on genetic information. It also prohibits such organizations and institutions from acquiring genetic information about employees or from requesting employees to undergo genetic testing. Genetic information may be disclosed when requested by the employee, when the request comes from an occupational or health researcher, in response to a court order, in response to a government official investigating compliance with GINA, or in connection with the employee's compliance with the Family and Medical Leave Act or a public health official (Code of Federal Regulations, Title 29—Labor).

GINA has, however, been criticized for providing insufficient protection for genetic information and for promoting "genetic exceptionalism," that is, treating genetic information as special and different from other medical information. Although GINA includes provisions to prevent discrimination in employment and health insurance rates and coverage, it does not protect individuals against discrimination in life insurance, disability insurance, or long-term care insurance (Erwin 2008, 871). Additionally, the protections against discrimination under GINA only hold for people with the genetic predisposition to disease or disability that has not yet been manifested. Once individuals become symptomatic, GINA does not protect them from discrimination (Rothstein 2008). Nor does GINA completely protect employees from their employers gaining access to genetic information. Employers can view the genetic information of employees who participate in a workplace "wellness program" and employees who provide "information voluntarily…or where the employee provides family history under the Family Medical Leave Act" (Erwin 2008, 871). At the same time, before an employee is hired, the employer may condition employment on the release of the future employee's complete medical record, which often includes some genetic information. Erwin points out that although it would be illegal for employers to discriminate against their employees based on genetic information gathered in these ways, this will be likely to continue to occur given that other forms of illegal discrimination continue, such as discrimination based on race, sex, religion, and so forth (Erwin 2008, 871).

Certificates of confidentiality are tools that researchers may use to help ensure the confidentiality of information about participants, including the fact that they participated in the research. Established by the Comprehensive Drug Abuse Prevention and Control Act of 1970 and modified in the Public Health Service Act, certificates of confidentiality were originally designed to encourage participation in studies that involved the participants revealing information about their illegal drug use (Currie 2005). The scope of protections that certificates of confidentiality offer was increased in 1974 by the Comprehensive Alcohol Abuse and Alcoholism Prevention, Treatment, and Rehabilitation Amendments, which allowed the certificates to cover "mental health research in general, including studies on use of alcohol and other psychoactive drugs" (Currie 2005, 8). Certificates of confidentiality now cover an even wider scope of research, as legislated in the Public Health Service Act, which includes "any research, whether funded by Department of Health and Human Services (DHHS) or not, where confidentiality is deemed essential for producing valid and reliable information" (Currie 2005, 8). Certificates of confidentiality offer protections: "for research collecting information that, if disclosed, could have adverse consequences or damage subjects' financial standing, employability, insurability, or reputation" (Beskow et al. 2008, 1054). The current law states that with a certificate of confidentiality, "persons engaged in biomedical, behavioral, clinical, or other research…may not be compelled in any Federal, State, or local civil, criminal, administrative, legislative, or other proceedings to identify such individuals" (Beskow, Dame, and Costello 2008, 1054). Certificates of confidentiality are meant to prevent researchers or their institutions from being forced to disclose individual research data or to identify research participants to law enforcement. Thus, certificates of confidentiality do not prevent information obtained in the course of a research project from being obtained by the participants' insurance companies and being used to deny coverage (Earley and Strong 1995; Kass 1993).

In practice, the courts have rarely tested certificates of confidentiality. So it is unclear how much protection they actually offer to research subjects. These certificates have come up in legal proceedings twice. In one court case, *People v. Newman*, 32 N.Y.2d 379 (1973), a certificate of confidentiality prevented the identity of someone in a drug treatment program from being released as part of a murder investigation (Beskow et al. 2008, 1054). The second court case involving a certificate was an appellate case, *State v. Bradley*, 179 N.C. App. 551 (2006). In this case the Duke University Health System requested an order for protection that would prevent information about a witness's participation in a mental health study from being used as exculpatory evidence in a criminal trial. The judge in this case was "unfamiliar with Certificates" (Beskow et al. 2008, 1054). Although the judge allowed the protective order and prevented the participant's information from being used by the defense, he did not mention the existence of the certificate of confidentiality as his basis for making the decision. Instead, he cited the fact that "the defense was unlikely to find exculpatory evidence" (Beskow et al. 2008, 1054).

This case offers ambiguous evidence for the effectiveness of the certificate because ultimately it was not the feature of the case that ultimately made the judge's decision. Because certificates are relatively untested in this way, some bioethicists and researchers are skeptical of their efficacy (Beskow et al. 2008; Currie 2005; Hermos and Spiro 2009; Melton 1990).

Data Sharing

Information obtained in the course of research can and ought to be shared between researchers. Sharing is useful because it facilitates research and the rapid increase of knowledge (Lowrance 2006, 4). Yet the way in which data are shared should be regulated to maintain confidentiality of the research participant's information. The demand to maintain confidentiality should be weighed against the potential social benefits of conducting research. This is especially true in cases where wide participation of subjects is necessary to make meaningful discoveries. The HMP is precisely that kind of research project. The NIH has a general policy that encourages sharing data as widely as possible while also maintaining the privacy of research participants in accordance with the Privacy Rule of HIPAA (National Institutes of Health 2003).

The NIH provides the following rationale for promoting and requiring data sharing:

> Sharing data reinforces open scientific inquiry, encourages diversity of analysis and opinion, promotes new research, makes possible the testing of new or alternative hypotheses and methods of analysis, supports studies on data collection methods and measurement, facilitates the education of new researchers, enables the exploration of topics not envisioned by the initial investigators, and permits the creation of new data sets when data from multiple sources are combined. By avoiding the duplication of expensive data collection activities, the NIH is able to support more investigators than it could if similar data had to be collected de novo by each applicant. (National Institutes of Health 2002)

There are a variety of ways in which data can be shared. Scientists may make the data truly public or make it available on databases that have built in protections of confidentiality. In some instances the NIH recognizes that some data sets cannot be widely shared, especially for studies with very small samples or those collecting particularly sensitive data. Such data may only be shared when stringent safeguards are in place to ensure confidentiality of the data (National Institutes of Health 2002).

Data can be modified in various ways to maintain confidentiality of the research participants.[13] Data can be *anonymous* or *de-identified*. The difference between *de-identified* and *anonymized* data is subject to debate. The key concept is whether a link can be established between one piece of information and another, more specifically whether a link could be established between the data obtained during research and the particular people who participated in the research. HIPAA specifies degrees to which data should be divorced from personal identifiers, such as social security numbers, medical record numbers, telephone numbers, and so forth (HIPAA §164.514(b)(2), n.d.).

Under the HIPAA Privacy Rule, data are de-identified if either (1) an expert determines that the risk that certain information could be used to identify an individual is "very small" and documents and justifies the determination, or (2) the data do not include any of 18 identifiers (of the individual or his relatives, household members, or employers) that could be used alone or in combination with other information to identify the subject: names, geographic subdivisions smaller than a state (including zip code), all elements of dates except year (unless the subject is older than 89 years old), telephone numbers, fax numbers, e-mail address, social security numbers, medical record numbers, health plan beneficiary numbers, account numbers, certificate/license numbers, vehicle identifiers including license plates, device identifiers

[13] Additional discussion of this issue can also be found in chapter 6.

and serial numbers, URLs, Internet protocol addresses, biometric identifiers, full face photos and comparable images, and any unique identifying number, characteristic, or code. Note that even when these identifiers are removed, the Privacy Rule states that information will be considered identifiable if the covered entity knows that the identity of the person may still be determined.

Alternatively, data could be *anonymized*; this refers to previously identifiable data that has been de-identified and for which a code or other link no longer exists. An investigator would not be able to link anonymized information back to a specific individual. Data could also be *anonymous*, which means that data were collected without identifiers and were never linked to an individual.

To illustrate the distinction between anonymous and de-identified data, we turn to a couple of cases. For example, in 2006 Netflix created a database of customers' movie rankings and published it on their site. Before publishing it they attempted to *anonymize* the information by coding each record with random numbers to replace names and removing any other data they deemed personal details such as reference to locations and so forth. Utilizing a database called the Internet Movie Database (IMDB), researchers at the University of Texas were able to *de-anonymize* the movie ratings posted by Netflix.

Another example that illustrates the difference between anonymized and de-identified data is disease monitoring. Data on human immunodeficiency virus (HIV)/acquired immunodeficiency syndrome (AIDS) may be de-identified and/or anonymized, but a possible linkage between the data and personal identifiers may still exist. Consider the case of an elderly Asian gentleman who resides in a small town in Minnesota and is a farmer. The entire population of the county is 1,000 people. He has been diagnosed with HIV that he contracted during a blood transfusion after a tractor accident. A researcher at a large research institution studying the spread of HIV decides that *county level* is an appropriate way to study the progression of the disease. The problem is that there are millions of people in some counties in the United States, while there are hundreds in others. When the results of the study are published, the data includes the race of the study participants. Given that there are only three elderly Asian farmers in that county, it is easy to deduce the identity of the HIV-positive elderly Asian male.

While this may seem to be an extreme case, the ability to link obscure data to other data is a real concern, and the HIPAA requirements for de-identification cannot completely eliminate the risk of reidentification when a public database is available.[14] Examples include Canadian researchers (Iqbal et al. 2010) who were able to reidentify anonymous e-mails by developing an algorithm that matched word usage patterns to establish the identity of individuals.

---

[14]It should be noted that the term *public* might be misleading. It could mean truly public, like a website, or it could refer to a database accessible only to researchers.

Similar to a fingerprint, they established a "write print" that identified patterns in word usage, case use, spelling habits, and so forth. Since this experiment was done on 158 individuals who worked for Enron, it is fairly easy to apply the "write print" to the dictation patterns (medical records) of physicians. As long as there is a link to a name somewhere, the potential for reidentification is real. Therefore, it is critical that truly anonymous records be carefully considered because for de-identified records the possibility of reidentification cannot be completely eliminated. These potential links can be very difficult to foresee when developing a data model. One should assume that there is always the potential to reidentify a record and that to truly render data anonymous, collaboration with a forensic data professional is required.

Although the ways in which confidentiality is currently protected have flaws, the general spirit of the NIH guidelines is congruent with our recommendations that data sharing among researchers should be fostered. The statements of the NIH on sharing data provide an accurate balance between fostering scientific progress and insistence on the protection of confidentiality for study participants. Many organizations already have terms of agreement, which restrict the release of data into the public domain.

We mentioned earlier that researchers share data in a variety of ways; some are public and others involve sharing in a much more controlled way. Typically, institutions allow data to be released under specific terms of agreement, which have the status of legally binding contracts between two interested parties. Some of the main elements of controlled-release terms of agreement require the commitment to prevent identifiability, including the requirement that data be de-identified. Terms of agreement might restrict the use of the data to only those who are planning to use it for particular types of research projects and might restrict further transfer of data. Terms of agreement for controlled release might include requirements for confidentiality protections, which require adherence to HIPAA, and might designate those who are responsible to ensure such adherence to regulations. Further, they might include the commitment that researchers will not attempt to identify and recontact the subjects in the original study (Lowrance 2006, 20).

It is difficult to predict the specific challenges that data from the HMP will pose to confidentiality protections, and it is hard to provide specific suggestions for how such protections ought to be adjusted to meet this challenge. The ethical justification for sharing data between researchers remains. Researchers, like physicians, ought to be appraised of their obligation to maintain confidentiality. This can be achieved through continued education in research ethics, with an emphasis on confidentiality. We should also adjust our notions of confidentiality to reflect the difference between sharing information publicly and sharing information for the purpose of research. To the extent that scientists' ethical commitment and legal obligation to maintain confidentiality is acknowledged and accepted, sharing data among scientists should not be perceived as a threat to confidentiality. We should not confound data sharing among scientists with unwarranted breaches of confidentiality.

We should, of course, not ignore the potential risks to individual participants from a breach of confidentiality and should strive to perfect ways in which confidentiality is protected. Finally, if scientific progress is a social good, we should accept that research on the HMP might pose some threats to informational privacy, but when those risks are kept minimal, they are justified by the social benefits of research.

Conclusion

In this chapter we explained how the HMP could produce novel ways of obtaining information about individual people as well as groups of people. We cited some contemporary studies that indicate that each individual's microbiome can be used to reveal details about a person's lifestyle, personal habits, nutrition, and health. These new ways to obtain identifying information could cause potential privacy infringements for research participants. In examining this possibility, we considered different conceptions of privacy and identified how the HMP could hypothetically produce infringements on informational privacy. We also examined possible ways in which such infringements could occur, but we tempered this worry by explaining that such privacy infringements are unlikely and that some risks to privacy are justified by the social benefits of research.

In this chapter, we also argued that the concept of confidentiality is a more suitable concept than privacy in the context of medicine and biomedical research. Considerations of privacy focus on how disclosure of private information should be limited or prevented, but both medicine and research require patients and subjects to disclose personal information under an expectation that the physician or scientist will safeguard it. Given that medicine and research sometimes require the disclosure of information, the focus should be on how to protect confidentiality and ensure that the standards for such protection already used in medicine are applied to research as well.

To that effect, we discussed existing legal protections of privacy and confidentiality including HIPAA and GINA. We also discussed the current guidelines that govern sharing of data and how those guidelines could apply to research on the human microbiome. We support the general spirit of those guidelines but caution that the research community should be continuously vigilant in their efforts to protect the confidentiality of study participants. As novel ways of obtaining individual information become available, and insofar as confidentiality remains socially valued, guidelines for protection of confidentiality should be adjusted to prevent foreseeable risks to the confidentiality of research participants.

Policy Recommendations

- All databanks collecting microbial samples should have protection from subpoena for criminal investigation.
- Insofar as research on the human microbiome could result in discrimination against particular individuals for medical insurance and employment, GINA-type legislation could be extended to apply to human microbial samples.
- Currently, data about the human microbiome are shared by request, and such requests are granted if the information is to be used for further research. Unless there are reasons to believe that data from a particular study are significantly more likely to result in identification of the participants, this practice ought to continue for the HMP.

References

Annas, George. 2003. HIPAA regulations: a new era of medical-record privacy? *New England Journal of Medicine* 348(15): 1486–1490.

Arumugam, Manimozhiyan, et al. 2011. Nature enterotypes of the human gut microbiome. *Nature* 473(May): 174–180.

Asch, Steven, Barbara Leake, and Lillian Gelberg. 1994. Does fear of immigration authorities deter tuberculosis patients from seeking care? *Western Journal of Medicine* 161(4): 373–376.

Beskow, Laura M., Lauren Dame, and E. Jane Costello. 2008. Certificates of confidentiality and compelled disclosure of data. *Science* 322(5904): 1054–1055. doi:10.1126/science.1164100.

Cheng, T. L., J. A. Savageau, A. L. Sattler, and T. G. DeWitt. 1993. Confidentiality in health care: a survey of knowledge, perceptions, and attitudes among high school students. *Journal of the American Medical Association* 269(11): 1404–1407.

Currie, Peter M. 2005. Balancing privacy protections with efficient research: institutional review boards and the use of certificates of confidentiality. *IRB: Ethics and Human Research* 27(5): 7–12.

Earley, C. L., and L. C. Strong. 1995. Certificates of confidentiality: a valuable tool for protecting genetic data. *American Journal of Human Genetics* 57(3): 727–731.

Erwin, Cheryl. 2008. Legal update: living with the Genetic Information Nondiscrimination Act. *Genetics in Medicine* 10(12): 869–873.

Fierer, Noah, et al. 2010. Forensic identification using skin bacterial communities. *Proceedings of the National Academy of Sciences of the United States of America* 107(14): 6477–6481.

Ford, C. A., S. G. Millstein, B. L. Halpern-Felsher, and C. E. Irwin, Jr. 1997. Influence of physician confidentiality assurances on adolescents: willingness to disclose information and seek future health care: a randomized controlled trial. *Journal of the American Medical Association* 278(12): 1029–1034.

Fortenberry, Dennis J., et al. 2010. Urethral microbiome of adolescent males. http://www.ncbi.nlm.nih.gov/projects/gap/cgi-bin/study.cgi?study_id=phs000259.v1.p1. Accessed June 21, 2012.

Francis, Leslie Pickering. 2008. Privacy and confidentiality: the importance of context. *Monist* 91(1): 52–67.

Fried, Charles. 1968. Privacy (a moral analysis). *Yale Law Journal* 77(1): 475–493.

Hermos, John A., and Avron Spiro III. 2009. Certificates should be retired. *Science* 323(5919): 1288–1289.

HIPAA §164.514(b)(2). n.d. Other requirements relating to uses and disclosures of protected health information. http://www.gpo.gov/fdsys/pkg/CFR-2002-title45-vol1/pdf/CFR-2002-title45-vol1-sec164-514.pdf. Accessed June 21, 2012.

Iqbal, Farkhund, et al. 2010. Mining Writeprints from anonymous e-mail for forensic investigation. *Digital Investigation* 7: 56–64.

Kass, N. E. 1993. Participation in pedigree studies and the risk of impeded access to health insurance. *IRB* 15: 7–10.

Lowrance, William W. 2006. Privacy, confidentiality and identifiability in genomic research. Discussion document for workshop convened by the National Human Genome Research Institute, Bethesda, MD, October 3–4, p. 4.

Melton, Gary B. 1990. Certificates of confidentiality under the Public Health Service Act: strong protection but not enough. *Violence and Victims* 5(1): 67–71.

Moore, Adam D. 2003. Privacy: its meaning and value. *American Philosophical Quarterly* 40(3): 215–227.

Nasidze, I., et al. 2009. Global diversity in the human salivary microbiome. *Genome Research* 4: 636–643.

National Institutes of Health. 2002. NIH announces draft statement on sharing research data. http://grants.nih.gov/grants/guide/notice-files/NOT-OD-02-035.html. Accessed June 21, 2012.

National Institutes of Health. 2003. Final NIH statement on sharing research data. http://grants.nih.gov/grants/policy/data_sharing. Accessed June 21, 2012.

Rachels, James. 1975. Why privacy is important. *Philosophy and Public Affairs* 4(4): 323–333.

Ravel, Jaques, et al. 2011. Vaginal microbiome of reproductive-age women. *Proceedings of the National Academy of Sciences of the United States of America* 108(Suppl. 1): 4680–4687.

Rothstein, Mark A. 2008. Putting the Genetic Information Nondiscrimination Act in context. *Genetics in Medicine* 10(9): 655–656.

Scanlon, Thomas. 1975. Thomson on privacy. *Philosophy and Public Affairs* 4(4): 315–322.

Schwab, Abraham, Lily Frank, and Nada Gligorov. 2011. Saying privacy, meaning confidentiality. *American Journal of Bioethics* 11(11): 44–45.

Solove, Daniel J. 2006. A taxonomy of privacy. *University of Pennsylvania Law Review* 154(3): 477–560.

Thomson, Judith Jarvis. 1975. The right to privacy. *Philosophy and Public Affairs* 4(4): 295–314.

5

Research Ethics

Rosamond Rhodes, Martin J. Blaser, Joseph W. Dauben, Lily E. Frank, Daniel A. Moros, Sean Philpott

Introduction

In chapter 1 we reviewed the basic physiology of multicellular organisms as well as our understanding of the relationship between the microbial world and the physical structure of the Earth's atmosphere and crust. Ultimately, this is part of our growing understanding of the interaction of microorganisms, the environment, and the biologic world.

Awareness of these interactions is not entirely news. For the past 100 years, high school biology courses have taught students how bacteria around the roots of plants convert the relatively inert and unavailable nitrogen of the atmosphere into ammonia (NH_3), a process referred to as nitrogen fixation. NH_3 is then further oxidized into nitrates, which are more readily taken up by the root system. The effectiveness of agriculture in today's world and our ability to maintain a population of more than seven billion humans is dependent on the development of the Haber-Bosch process.[1] Without the development of this process of deliberately supplementing the activity of the microbial world, there would not be sufficient nitrogen fixation to maintain today's agriculture practice (Erisman 2008).

Microbiome research has been a part of medicine since long before Joshua Lederberg coined the term *microbiome* in 2001 to designate the combination of an organism and its microbial community (Lederberg and McCray 2001). As early as 1885, biomedical researchers began to recognize that the asymptomatic carriers of *Salmonella typhi* and *Corynebacterium diphtheria* were the critical

[1] This is the commercial synthesis for converting atmospheric N_2 into ammonia. Fritz Haber and Carl Bosch were each awarded the Nobel Prize in 1918 and 1931, respectively, for their work developing the process.

link to understanding the appearance and spread of diphtheria and typhoid fever. In the late nineteenth century, *Salmonella* species (not always *S. typhi*), and to a lesser extent *C. diphtheria,* were major contributors to the more than 20% mortality rate in children between birth and five years of age in the tenements and slums of most industrial urban centers. Much of this early work on asymptomatic carriers was conducted under the leadership of Hermann Biggs and William Park of the New York City Health Department. Park developed a culture kit that allowed diphtheria cultures to be taken in the home of potentially infected individuals (Park 1931). Cultures taken from household and tenement contacts of people with diphtheria allowed researchers to develop data on the role of asymptomatic carriers in the spread of the disease. This development was valuable on two fronts. First, it allowed for the effective use of an antitoxin that was most effective when applied early in the course of the disease. Second, this study highlighted the role of microscopic organisms in the spread and cure of infectious diseases.

As research on the human microbiome increases, it is important to avoid the mistakes of the past and navigate the dilemmas that accompany new technology and models of human subject research. Human microbiome research will involve huge numbers of human subjects and generate voluminous quantities of data to advance our understanding of the human microbiome and to use what we learn about it to promote health and alleviate disease. Of course, biomedical research involving human subjects should uphold the highest ethical standards. But do we know what those standards are? Starting with the Nuremberg Code, the Declaration of Helsinki, and the Belmont Report, up until and including the most recent version of the U.S. Common Rule (Code of Federal Regulations, Title 45, Part 46 [CFR 45 Part 46] 1991), standards for the ethical conduct of human subject research have largely developed in response to the revelation of serious ethical transgressions in human subject research. This history of reactive development to crises is prone to both exaggerated responses and lacunae.

The Nuremberg Code (1949) developed in response to the atrocities perpetrated by Nazi researchers on concentration camp inmates, and the Belmont Report (1979) developed in response to revelations of serious abuses of human subjects in studies within the United States, including the Tuskegee syphilis study and the Willowbrook hepatitis study. These studies were unethical because they were performed on subjects who were taken to be less worthy of respect and consideration than other individuals in the community, and therefore, subjects were treated in ways that no other person should be treated. In the case of Nazi research, the mistreated test subjects were the Jews, Catholics, Communists, gays, among others. In the United States, those whose worth was discounted included African American men, mentally disabled children, prisoners, and the unbefriended elderly. These research subjects were treated in ways that demonstrated an implicit assumption that such individuals could be used to create benefits for others, presumably because they had little value in the eyes of researchers and those responsible for research oversight.

In contrast, the envisioned human microbiome studies will aim to include samples from people of every sort, with the worth of every group acknowledged by its inclusion. Furthermore, the horrific research studies that served as touchstones for those who developed today's research regulation involved the deliberate infliction of significant harm, unreasonable burdens that no one who had a choice would accept under the circumstance. In contrast, the sampling involved in mapping the human microbiome will, for the most part, require just swabs and stool samples and, therein, involve no physical harm.

To prevent future scientific misconduct, successive documents have developed rules and principles designed to prevent unacceptable treatment of human subjects. In that light, our current standards for the ethical conduct of human subject research have been clarified, modified, and revised. On July 22, 2011, the U.S. Department of Health and Human Services (HHS) initiated a new effort to improve the rules that govern federally funded research. It posted an advance notice of proposed rulemaking (ANPR), "Human Subjects Research Protections: Enhancing Protections for Research Subjects and Reducing Burden, Delay, and Ambiguity for Investigators." In this chapter, we join this evolutionary process by examining the reigning standards of research ethics from the perspective of human microbiome research. We use this opportunity to take a fresh and critical look at the current research regulations and guidance. As suggested by the HHS notice, our aim is to assess the appropriateness and adequacy of the rules.

The Landscape of Microbiome Research

At least three different types of microbiome research are already under way. They all study human subjects, although each kind involves a different degree of risks and burdens. The first model involves the collection of a large number of samples from a broad spectrum of subjects so that scientists can answer very general questions. The second model examines individuals who have or do not have a particular disease to understand the role of microbiota in the development of specific diseases. The third employs probiotics or bacteriophages as clinical trial interventions to develop treatments for the possible cure or amelioration of the symptoms of specific diseases (Ventura et al. 2011). Probiotics are microbes used to modify the microbiome and provide a health benefit. Bacteriophages, or phages, are naturally occurring viruses that may be used to modify the microbiome by eliminating specific pathological bacteria or by reducing excessive proliferation of particular kinds of bacteria to maintain a healthy dynamic equilibrium. The lytic enzyme produced by a phage is a lysin. A lysin "degrades the bacterial cell wall, causing the bacteria to explode" (Borrell 2012).

The ultimate goal of the first kind of study is to understand how microbial communities are structured and how they function. These studies aim at answering questions such as: What sorts of bacteria, viruses, and fungi comprise the human microbiome? How many kinds of microbiota are common to all humans? To answer these questions, scientists must first assemble large collections of samples from human subjects and measure the extent to which they vary from individual to individual. We anticipate that the first kind of microbiome research will involve two significant features. These features are shared with genetic research involving biobanked samples.

First, samples from multiple sites on a very large number of individuals will be needed to develop results that are relevant to a broad spectrum of humans. Storing and processing samples, careful record keeping, extracting genetic material from samples, running genetic sequencing, and generating useful data is a complex, time-consuming, and very costly process. For this investment to have a significant payoff, data have to be widely available to researchers so that they may be used in many studies. To facilitate the efficient use of the microbiome samples and data, the National Institutes of Health (NIH) is creating federal repositories so that they can be widely shared among researchers. Study participants donate samples hoping that the collected microbiome samples will contribute to advancing biomedical knowledge. Second, the vast majority of these studies will expose participants to only very minimal risks of physical harm. Together, these features are significant because they provide a very different perspective for considering research ethics. Instead of looking backward at harms perpetrated by scientists in the past and aiming to avoid such problems in the future, the new microbiome vantage point directs us to take note of future benefits from new models of research that existing rules had neither imagined nor considered.

For example, Bruce Birren's study, "Defining the Human Microbiome," involved collecting and sharing large numbers of biological samples from human subjects. He explained:

> We will take advantage of the explosive advances in new sequencing technologies to rapidly create a catalog of over 400 reference microbial genome sequences, including those of bacteria, eukaryotes, and viruses, as well as generate extensive community profiles from a number of human body sites with a focus on the oral and vaginal microbiome. We will also aggressively develop and apply new methods and technologies to further reduce the cost and expand the reach of the HMP. These data will be rapidly released to the public and, along with those of our collaborating partners in the HMP network, will define the Human Microbiome communities and provide baseline measurements for a wealth of subsequent experimental research. (Birren 2011)

Another study by Koenig et al., "Succession of Microbial Consortia in the Developing Infant Gut Microbiome," aimed at identifying how the infant gut microbiome develops. This 30-month study collected fecal samples from the diapers of healthy infants along with a diary of their diet and health status. The data contributed to our understanding of how the human infant gut microbiome is assembled, and how life events relate to the microbiome's composition and function (Koenig et al. 2010).

In both of these studies, large numbers of samples were needed, whereas the physical risks to study participants were limited to the minute risks associated with collecting oral and vaginal swabs in one study, and the nonexistent risk of using fecal samples from the diapers of healthy infants in the other. The current standard for describing low-risk research is "minimal risk." The Code of Federal Regulations, 45 CFR 46.102, defines "minimal risk" research as a study in which "the probability and magnitude of harm or discomfort anticipated in the research are not greater in and of themselves than those ordinarily encountered in daily life or during the performance of routine physical or psychological examinations or tests." Given the increasing use of genetic and microbiome sample studies that involve only minute physical risk, and how these risks seem to be significantly different from other research risks, we see a need for distinguishing such studies from the typical clinical trials. Clinical trials typically involve testing a new, potentially harmful intervention on subjects to determine its safety and efficacy. We are introducing the term *de minimis risk* to mark off those minimal-risk studies with an exceedingly low degree of risk. A *de minimis* risk is a negligible risk that involves only hard-to-imagine and extremely unlikely physical risks to the body. The term *de minimis risk* would apply only to those minimal-risk studies with this extremely low level of risk, no more, say, than what is involved in taking a cheek swab.

The second human microbiome research model will resemble current research that aims at developing our understanding of disease processes. It will require the participation of individuals with the target disease as well as the participation of individuals who are free of the target disease to serve as a control group. Researchers will try to identify the microbiota associated with the disease by comparing the site-specific microbial samples from individuals who are not affected by the disease with samples from those who are. This may help them to determine whether microbiome differences are causes or effects of the target condition.

An example of this type of study is "The Vaginal Microbiome: Disease, Genetics and the Environment," by Fettweis et al. (2010). It aimed at providing an analysis of the vaginal microbiomes of normal women of reproductive age and women in the same age range with common pathological conditions (e.g., vaginosis, vaginitis, viral infections, bacterial sexually transmitted diseases). It studied samples from women in three different cities and from three different ethnic/racial groups: European Caucasian, African American, and Mexican

Hispanic. Again, the study sought to identify differences in the microbiome that had some bearing on pathology. Again, the vaginal smear to collect samples involved only *de minimis* risks (Fettweis et al., 2010).

Participants in this kind of microbiome research will be chosen because they either have the target condition or not, and not because we value them less than others. Again, most of the microbe sampling involved will require only that subjects provide swabs and stool samples and, hence, involve no physical harm. Studies that require samples from genitalia (like the one described earlier) or samples of stool may involve some embarrassment or unpleasantness, but nothing more enduring or substantial than that short-lived aesthetic or social discomfort. Studies that require specimens from the gut are more invasive and involve some degree of risk. When a small extra sample is taken during a scheduled diagnostic procedure, however, very little additional risk is involved. These risks to subjects will have to be minimized and justified as in other clinical research. Yet, such studies may be essential for gaining an understanding of the role of microorganisms in chronic debilitating autoimmune diseases and inflammatory diseases.

The third type of study will be investigations of the effectiveness of probiotics, phages, and lysins. According to the World Health Organization definition, probiotics are "live microorganisms which when administered in adequate amounts confer a health benefit on the host" (World Health Organization 2003). Research on the effectiveness of probiotics will, in some respects, resemble research involving infectious disease, and, in other respects, resemble drug development research.

Although the typical clinical trial of a biomedical intervention affects only the study participants, trials involving infectious disease, or even vaccines against infectious disease, can affect others. Because the microbiome is, to some extent, communicated between individuals, studied probiotics may communicate between individuals and may have untoward effects. For example, could a clinical study using probiotics as a treatment for necrotizing enterocolitis in premature infants in a hospital nursery perhaps cause some untoward effect on other infants in the facility? Presently, it is thought that probiotics do not remain within an individual for very long (not more than a few days), that they will not survive outside of the body, and that there will not be enough of them to communicate any effect to others. Some probiotics, however, may be more resilient than others, and in the future, particularly after genetic engineering to increase their longevity, they may be more communicable than they are now. In addition, we already know that there is tremendous variety among healthy people. Thus, bacteria that are benign in some individuals may be toxic in others (e.g., *Helicobacter pylori* causing stomach ulcers in some individuals) (Blaser 2012).

Bacteriophages (i.e., phages), in themselves, are not harmful to the cells of animals or plants, and they have not been found to produce side effects in humans. Furthermore, because each phage targets a specific bacterium, using them as therapy is less likely to disrupt the balance of the entire microbiome than would broad-spectrum antibiotics. Thus, it is reasonable to expect that the risks involved in human subject research of phage therapy will be minimal (Borrell 2012).

As bacteria evolve and become antibiotic resistant, the infections that they cause become serious problems, particularly within hospitals. For example, the risk of developing a methicillin-resistant *Staphylococcus aureus* infection, also known as MRSA, is a danger for patients who undergo surgery. Using phages or lysins, the enzymes produced by the viruses, to eradicate the pathological bacteria is, therefore, an attractive option (Rescha, Moreillon, and Fischettia 2011) that has been used with success in the former Soviet Union (Osborne 2000). For example, the first clinical trial of a lysin to specifically combat MRSA is scheduled to begin this year (Borrell 2012).

The environmental impact of microbial products will have to be assessed, regardless of whether they are classified as drugs (i.e., biotherapeutics) or as food (i.e., probiotics). Probiotics are not inert; they are live, mutable, and potentially self-replicating organisms that may interact in unpredictable ways with the normal microflora of the human body. Because these living organisms can inhabit the gut and interact with other gut flora, there is a possibility that drug-resistant strains can develop. To prevent antibiotic resistance from being communicated to other microbes or to humans, antimicrobial susceptibility testing will have to be incorporated into the development and manufacture of probiotics.

Similar possibilities of environmental impact may be associated with bacteriophage viruses. Phages can significantly affect microbial communities by destroying specific bacteria and thereby affecting the dynamic equilibrium of the microbiome. Also, as they destroy bacteria, bacterial by-products, which may have some harmful effects on the surrounding environment, are released (Osborne 2000). One study reported to have identified 991 phages and over 130 bacterial hosts (Stern et al. 2012). This finding suggests that there is significant potential for the use of phages as novel anti-infective agents. Using current phage therapy techniques, a number of different phages are needed to treat each infection, and each treatment has to be tailored to the specific bacteria in the individual patient. Although no bacteria have been found to be resistant to lysin enzymes, phage-resistant bacteria evolve as do phages. When phage therapy is specifically targeted to individuals, the likelihood of developing phage resistance is minimized because of the specificity of the treatment. The mutability of both bacteria and phages, however, means that the dynamic process will require ongoing attention.

Lytic enzyme therapy is different from phage therapy. Newer techniques are using genetic sequencing of bacteria to determine which lytic enzymes (or lysins) are likely to be effective in destroying a particular bacterial strain and then employing the lysins directly as treatment (Schmitz et al. 2011). This technology could be more efficient and effective than previous phage therapy and perhaps also less susceptible to problems of creating treatment resistance. Lysins can be injected into the blood or applied directly to the skin or mucous membranes. In laboratory studies lysins have not been shown to develop resistance, but measures should be developed to monitor and prevent the development of bacterial resistance (Fischetti 2006).

For all of these reasons, clinician researchers and research oversight agents will have to consider the possible untoward impact on others and the environment when they undertake studies that introduce microbes as treatment for some diseases. They will have to determine whether the risk of harm to those who are not participating justifies imposing additional burdens (e.g., quarantine) on those who are. In some circumstances, at least until the safety of a new microbial intervention can be assessed, such unusual research measures may be ethically required.

Drug development research requires clinical data to demonstrate both the safety of the product and its efficacy in the treatment of a medical condition or disease. Animal and human studies in support of drug approval must conform to federal standards for the ethical treatment of animals and the ethical conduct of human subject research. In the United States, the Food and Drug Administration (FDA) oversees the approval and manufacture of new drugs to ensure the consistency and quality of the products. Securing approval of a drug for a specific condition involves a lengthy and expensive process to meet rigorous evidence standards. The standards do, however, provide physicians who write prescriptions and patients who use the drugs with confidence in their quality, consistency, safety, and efficacy when used according to recommendations.

Increasing numbers of probiotics are currently being recommended by physicians (Cordina 2011) and sold directly to consumers, but very few are being sold as drugs. For the most part, consumers who purchase and use probiotics believe that they are using a product that promotes health or alleviates disease or symptoms. Actually, the probiotics that are marketed as food, food additives, or dietary supplements have not been proven effective with a standard of evidence that is scientifically compelling.

As we explained, there are three very different roles for microbiome research: (1) learning about the human microbiome, (2) studying the role of microbiota in disease susceptibility and resistance, and (3) testing probiotics and phages as treatments. Each raises very different kinds of ethical considerations that may require different research practices. There is no reason to presume that the very same shoe will fit every foot, or that the very same rules should govern every kind of research. This chapter will, therefore, examine the current standards for research ethics in the context of microbiome research and, in that light, offer recommendations for making changes that are sensitive to these different kinds of research. We shall begin with a review of some of the relevant history of research ethics and some of the foundational commitments of current policy.

Historical Development of Research Ethics Regulation and Guidelines

We next present a condensed review of the relevant history of research ethics to illustrate that promulgating codes and regulations for human subject research is a relatively recent phenomenon. The current codes, guidelines, and philosophical work on the ethics of human subject research represent an evolving train of thought rather than immutable commandments. Furthermore, for the most part, the current regulations represent a narrow focus on historical examples of research misconduct and dangerous clinical research. Thus, it is time to revise the regulations in response to new scientific developments, such as the Human Microbiome Project (HMP); emerging technological developments; and shifting research paradigms.

Early concern about the ethical conduct of human subject research was expressed by some physicians, such as the Scottish physician-ethicist John Gregory (1724–1773) and the English physician Thomas Percival (1740–1804). Percival explicitly addressed the importance of sympathy and "fellow feeling" as guidance for limiting the risks and burdens imposed on research subjects (Haakonssen 1997, 72; Percival [1803] 1985). Notably, Claude Bernard, French physiologist and father of experimental medical research, developed a comprehensive and systematic set of ethical requirements for the conduct of animal and human subject research. In his work on the scientific method, *An Introduction to the Study of Experimental Medicine* (1865), he offered ethical standards for studies involving human subjects. There he insisted that experiments "that can only harm are forbidden. Those that are innocent are permissible, and those that may do good are obligatory" (1865/1957, 102). Even though these texts were published, it is not clear that the ideas expressed in them were generally acknowledged or broadly shared.

Nevertheless, toward the end of the nineteenth century, research advances due to new discoveries in pharmacology, bacteriology, and immunology, along with the development of new equipment and technologies like the x-ray, greatly expanded the use of human subjects in medical research. These developments were reflected in the founding of new journals such as the *Journal of Experimental Medicine* (1896), the *Journal of Medical Research* (1896), and the *American Journal of Physiology* (1898). Medical practitioners experimented on themselves. A considerable amount of experimentation in the late nineteenth and early twentieth centuries was also conducted on animals, a practice that led to vigorous antivivisectionist campaigns. Because some diseases did not have suitable animal research models, physicians turned instead to human subjects for the study of diseases like syphilis and gonorrhea. The most commonly used subjects for medical research came from marginal groups: prostitutes, prisoners, mentally ill, terminally ill, children, slaves, and members of the armed services. For example, in 1895 New York pediatrician Henry Heiman described successful infection with gonorrhea of a 4-year-old boy, ("an idiot with chronic epilepsy"), a 16-year-old boy (an "idiot"), and a 26-year-old man in the final stages of tuberculosis. Vaccine testing was another area in which children were natural subjects because they were not previously exposed to the target. In 1895 Walter Reed and George Sternberg studied various vaccines for smallpox immunity by vaccinating children at several orphanages in Brooklyn, New York. Similar trials were undertaken with vaccines for measles, chickenpox, scarlet fever and tuberculosis (Lederer and Grodin 1994).

Not all studies were performed in conformity with reigning standards. For example, on December 29, 1900, the Prussian government issued a directive in response to the case of Dr. Albert Neisser, who had studied immunization against syphilis by inoculating healthy subjects with serum from syphilitic patients. Almost all of the subjects of his experiments were prostitutes, and all contracted syphilis. None had been informed about how Dr. Neisser was using them, none were warned about the likely consequences of the research, and none had consented to their role. The Prussian regulations published in the wake of these revelations prohibited human experimentation on minors and the mentally incompetent. They also required informed consent from research subjects as well as full disclosure of any possible adverse consequences that might result from the experiments. Still researchers did not always adhere to this directive. Later guidelines had to be issued by the German Reich in 1931 after the deaths of 75 children in Lübeck following trials of a tuberculosis vaccine.

Atrocities conducted by Nazi researchers during World War II led to the Nuremberg Code of medical ethics, as formulated by American judges during the "Doctors' Trial" in Nuremberg, Germany. There, 23 physicians and scientists stood accused of murder and torture as a result of their "medical research" in Nazi concentration camps. The Nuremberg Code was issued as the judges handed down their decision on August 19, 1947.

In *Subjected to Science: Human Experimentation in America Before the Second World War*, historian Susan Lederer shows that in the United States "even before 1930, researchers observed limits in their experiments with human subjects. Although lacking enforcement policies and far from perfect, ethical guidelines influenced the conduct of research with both human and animal subjects in the decades before World War II" (Lederer 1995, 17). Yet, several decades after the Nuremberg trial, concern for research involving human subjects was brought to national attention by Dr. Henry K. Beecher's exposé of medical research in his paper "Ethics and Clinical Research," published in the *New England Journal of Medicine* (1966). Beecher was particularly critical of lapses in informed consent and cited more than 30 examples of research published in peer-reviewed scientific journals that appeared to violate minimum standards.

U.S. studies that came to the attention of the wider public in the 1960s were especially troubling. For example, studies were conducted at the Willowbrook State School in Staten Island, New York, where hundreds of mentally retarded children were infected with hepatitis to study the disease's natural development and the development of immunity. Another disturbing study was conducted at the Jewish Chronic Disease Hospital in Brooklyn, New York, in 1963. Live cancer cells were injected into 22 chronically ill, debilitated, elderly, noncancer patients to investigate the human transplant rejection process. Also around this time, the infamous Tuskegee Syphilis Study was brought to the public's attention. Beginning in 1932 the U.S. Public Health Service headquartered a national syphilis study group at the Tuskegee Institute (later Tuskegee University). This study involved observation of the natural progress of the disease in nearly 400 infected patients, all poor black men from Macon County, Alabama. These men were never told that they had syphilis. Over the four decades the study lasted, they were never offered treatment, not even when penicillin became the standard remedy for the disease in the late 1940s (Reverby 2009). Although ethical concerns about the Tuskegee study were raised as early as 1966, it was not until front-page stories appeared in the *Washington Star* and the *New York Times* in July of 1972 that the research was stopped. Congressional hearings eventually led to the appointment of a National Commission for the Protection of Human Subjects of Biomedical and Behavioral Research in 1974. In May of 1974, passage of the National Research Act established regulations calling for the creation of institutional review boards (IRBs) for all research conducted by the U.S. Department of Health, Education, and Welfare (HEW), which has now been split into the Department of Health and Human Services (DHHS) and the Department of Education (ED).

In the same period, concern about the need for informed consent in research was growing. Prior to World War II, there were only a few isolated examples of researchers who elicited informed consent from study participants. As James H. Jones (1993) maintains in his history of the Tuskegee Syphilis Study, in the United States, and Susan Lederer in hers, there was "no system of normative ethics of human experimentation that compelled medical researchers to temper their scientific curiosity with respect for the patient's rights" (Lederer, 1995, p. xv). Lawsuits, however, were the catalyst for promoting concern for informed consent. According to Lederer, "By the early 20th century, growing numbers of lawsuits over unauthorized surgical procedures prompted surgeons and hospital administrators to formalize the process of obtaining patient consent" (Lederer 1995, 16). Similarly, as Hornsby and Schmidt observed in their study of *The Modern Hospital* (1913), the growing fear was that "...some venal lawyer looking for a fee happens to gain the ear of the family of the patient" (Hornsby and Schmidt 1913, p. 46).

Critics of research practices also raised questions about the propriety of providing direct payment or other benefits for participation in medical research (Lederer 1995, p. 123). For example, in 1908 the government paid subjects $25 to $30 per month to participate in experiments to determine the safety of saccharin. In 1911, after the federal government challenged the Coca-Cola Company's advertising claim to being "the ideal brain tonic," Coca-Cola paid subjects for their participation in a study to ascertain the extent of stimulation caused by ingesting differing amounts and concentrations of caffeine. During the Depression, employment bureaus sent jobless men to laboratories to serve as paid research subjects. Those who agreed to take part in the infamous Tuskegee syphilis study were offered hot meals, transportation, and $50 for permission to allow postmortem dissections. In 1958 Tuskegee survivors were given a certificate and $1 for every year they had participated in the study. In a 1978 Belmont Report, the National Commission for the Protection of Human Subjects of Biomedical and Behavioral Research recommended "Limiting remuneration to payment for time and inconvenience of participation and compensation for any injury resulting from participation..." (1979, 25). Today, virtually all codes of research ethics prohibit "undue inducement" because of its presumed interference with consent.

Another concern that figured into research ethics regulations focused on the populations recruited to participate in studies. Philosopher Hans Jonas was among the earliest to raise concern about protecting the vulnerable from research. In his widely read article "Philosophical Reflections on Experimenting with Human Subjects," published in *Daedalus* in 1969, he argued:

> The poorer in knowledge, motivation, and freedom of decision (and that, alas, means the more readily available in terms of numbers and possible manipulation), the more sparingly and indeed reluctantly should the reservoir be used, the more compelling must therefore become the countervailing justification.... It will mean that [researchers] will have to fight a strong temptation to go by routine to the readiest sources of supply–the suggestible, the ignorant, the dependent, the 'captive' in various senses. (Jonas 1969, 237)

Within five years of the passage of the 1974 National Research Act, under the auspices of the HEW, the Belmont Report[2] went further. It promulgated a national set of "Ethical Principles and Guidelines for the Protection of Human

[2] The Belmont Report was named after the Belmont Conference Center where the report was first issued in 1978.

Subjects of Research" (1979). The Belmont Report was strongly influenced by Han Jonas's paper on the use of human subjects in medical research. It stipulated requirements of respect for persons, informed consent, beneficence, and justice.

Research Ethics and Human Microbiome Research

The Federal Policy for the Protection of Human Subjects produced in 1991, now referred to as The Common Rule, has served as the framework for guiding human subjects research. Yet, from the start of the Human Genome Project (HGP) some people have recognized that this framework needs to be adapted to take account of emerging technologies and changing models of research. In fact, the agencies of the U.S. government overseeing the HGP have set aside 3% to 5% of their annual HGP budgets to support studies of the ethical, legal, and social implications (ELSI) of genetics research and studies that raise questions about the availability, distribution, and use of genetic information. Today, the U.S. government funds the world's largest program addressing societal issues related to genomics research (http://www.ornl.gov/sci/techresources/human_Genome/elsi/elsi.shtml). Thus far, ELSI studies have addressed questions about privacy and confidentiality and how genetic information might be used by insurers, employers, courts, schools, adoption agencies, and the military, among others; ownership and control of genetic information; possible psychological repercussions for subjects and possible stigmatization due to genetic factors, especially subjects in minority or other marginalized communities; and reproductive issues including proper counseling about risks and limitations of genetic technology, including fetal genetic testing. In addition, ELSI research is directed to the researchers and institutions that either carry out research or oversee and manage the results of that research, including the massive databases that are being collected in conjunction with a wide variety of genome-related studies. All of these historically significant ethical concerns are also related to the human microbiome research (Foster 2004; McGuire 2010).

The HMP, a development of the HGP, is just one of several similar initiatives around the world. These include the Human Epigenome Project, to identify, catalog, and interpret genome-wide DNA methylation patterns of all human genes in all major tissues, with direct links to cancer research; the Cancer Genome Project, to identify sequence variants and mutations critical to the development of human cancers; the UK Biobanks project (with up to 500,000 participants), to study the roles of nature and nurture in health and disease; the Estonian Genome Project Foundation (EGPF); the Norwegian Mother and Child Cohort Study (involving 100,000 mothers, 100,000 children, and 70,000 fathers); the International HapMap project, to develop a haplotype map of the human genome, which will describe the common patterns of human genetic variation and is expected to help find genetic variants affecting health, disease, and response to drugs and environmental factors, all of which have direct relevance to the microbiome; and the Human Variome Project, to collect all human genetic variations affecting human health.

One of the most controversial of these projects is deCODE, a private venture in Iceland that seeks to use population resources to convert research on the genetic causes of common diseases into a range of products and services in gene and drug discovery, DNA-based diagnostics, pharmacogenomics, and bioinformatics. All of these activities build upon deCODE's proprietary research in human genetics as well as contract service work for pharmaceutical and biotechnology industries. Since its inception, some people have raised concerns about potential bias in their research oversight (Abbott 1999).[3]

Others have raised concerns about the Genographic Project, a cooperative venture of the National Geographic Society and IBM, which maps historical human migration by collecting and analyzing DNA samples from hundreds of thousands of people in all parts of the world. This project illustrates how controversial genetic-related research can be. No sooner had this undertaking been announced in 2005 than the Indigenous Peoples Council on Biocolonialism (IPCB) called for a boycott of the project, and in May of 2006 the United Nations Permanent Forum on Indigenous Issues (UNPFII) recommended the project be suspended. Some feared land rights might be endangered; others were concerned that the results might contradict indigenous beliefs about ancestral origins. Some Indian tribes in North America have refused to take part, claiming that "what the scientists are trying to prove is that we're the same as the Pilgrims except we came over several thousand years before" (Harmon 2006).[4] The Genographic Project is, however, being supported by many groups around the world. Its success is closely linked with advances in technology and computer analysis of data that even a decade ago would have been virtually unmanageable.[5]

Environmental factors that vary widely from community to community, and from one part of the world to another, can affect both the human genome and the human microbiome. Thus, it is important to study a variety of communities across the globe. Questions about what the ethical conduct of genetic research entails are also relevant to the HMP. We should be mindful of the

[3] For a reply see Gulcher and Stefansson (1999); for details about the company's recent emergence from bankruptcy and abandonment of their drug development work, see Wade (2010); see also Rose (2003), which examines the ethical issues the Swedish UmanGenomics has faced where irreconcilable questions about rights to access and ownership of the database have arisen.

[4] Maurice Foxx, chairman of the Massachusetts Commission on Indian Affairs and a member of the Mashpee Wampanoag, quoted in Harmon (2006).

[5] For details, see Loughran (2005) and Zalloua (2008); for concerns about informed consent issues when research involves groups or entire human populations, see Greely (2001).

concerns raised by people around the globe when formulating and justifying research policies relating to both genetic and human microbiome research.

Critical Reflections on the U.S. Framework for Human Subject Research

In a 2000 article that appeared in the *Journal of the American Medical Association*, "What Makes Clinical Research Ethical?" Ezekiel Emanuel, David Wendler, and Christine Grady of the NIH briefly summarize the seven principles that they take to be the conceptual framework of research ethics. More recently, a simplified version of that paper, "Research Ethics: How to Treat People Who Participate in Research," by Emanuel, Emily Abdoler, and Leanne Stunkel, was posted as a brochure on the NIH website (http://www.bioethics.nih.gov/education/FNIH_BioethicsBrochure_WEB.PDF). We shall employ their seven principles framework as a template for commenting on the ethical conduct of research with human subjects. First, we shall offer our critical reflections on today's ethical framework for research ethics by discussing each of the principles in turn. Then we shall discuss how those seven principles apply to human microbiome research.

The first two principles offered in these documents are critically important for the ethical conduct of human subject research. They are, however, relatively uncontroversial. ***1. Value:*** Unless a study can produce knowledge or an enhancement of health or well-being, the use of resources and the exposure of human subjects to risks cannot be justified. ***2. Scientific Validity:*** Without validity, no knowledge can be produced. Thus, unsound research wastes resources and exposes subjects to risk without a justifying purpose. It is uncontroversial that without the promise of value and scientific validity, a study cannot be morally acceptable. Some of the subsequent principles raise more ethically complex issues.

3. Fair Subject Selection: Issues raised by this principle are particularly controversial and require lengthy discussion. The fundamental question is, who should participate in research? While everyone would agree that subject selection should be fair, what that entails is not entirely obvious. Emanuel et al. maintain that recruitment decisions should be based on scientific reasons and that exclusion decisions should be based on factors that would make some subjects especially vulnerable to risks. They explain that the "scientific goals of the study" should "be the primary basis for determining the groups and individuals that will be recruited and enrolled" (Emanuel, 2000, 2704). They go on to say that if a particular intervention is to be used in the treatment of a group, that group needs to be studied so that researchers can learn how it affects them. Furthermore, they hold that "those who may benefit should share some of the risks and burdens" (2705). Although these claims seem fair and reasonable to those who are not enmeshed in human subject research oversight, they are actually quite controversial.

The attitudes and practices that now govern human subject research are based on the sense that research participation is optional or supererogatory. This means that no one "should" participate in research. Even when someone hopes or expects to be a beneficiary of other people's study participation sacrifices, no one is to feel as if her own participation is morally required. The reigning view is that no one should in any way do or say anything that might make an individual feel any duty to participate.

According to this reigning view, biomedical research is an unusual activity that is, and should be, distinguished from clinical practice. This position stems largely from the previously mentioned influential 1969 paper, "Philosophical Reflections on Experimenting with Human Subjects," by Hans Jonas, who was writing with an eye still focused on Nazi atrocities. At the start of his article, Jonas expresses humility, given his status as a layman who does not have deep familiarity with medicine and biomedical research. Yet, ironically, Jonas goes on to make numerous assertions that express great confidence in his own understanding of the issues and his certainty about his conclusions. He asserts that the public interest can make its claim that individuals should participate as subjects in research only "where the whole condition of society is critically affected" (228). He goes on to hold that participation in biomedical research is a matter of "personal idealism—a matter of grace and not decree" (229), and that "[u]nless the present state is intolerable, the melioristic goal [of benefitting others] is gratuitous" (230). Jonas further maintains that, if anyone, scientist researchers, more than anyone else, should participate in research because "the higher the degree of the understanding regarding the purpose and the technique, the more valid becomes the endorsement of the will." (236) Jonas invokes the concept of *noblesse oblige* in arguing for what he calls the rule of the "descending order of permissibility" (235), according to which those with the greatest degree of understanding are to be the first participants in studies and those with the least understanding of the projects and their goals the last. He concludes by asking, "Who is conscriptable?" (237) He answers, "Least and last of all the sick" (238). He allows merely that "[p]atients should be experimented upon ... *only* with reference to *their disease*" (241).

Jonas's article is about "experimenting with human subjects," and he talks of individuals "sacrificing" themselves for the sake of "society" with references to death and conscription of youth in wartime. Perhaps he viewed research only in terms of "doing experiments on humans" where significant pain, suffering, and even death are inflicted upon subjects. This could have led him to conceptualize research ethics in a distinctly limited way. For example, he speaks of experimentation on subjects rather than systematic learning from the experiences of doctors and patients. Jonas describes individuals making sacrifices for the benefit of some abstract society, rather than seeing us all as cooperating in a collective enterprise to produce benefits for one another. He envisions clinical practice as a patient–physician dyad in which the physician is totally focused on the well-being of the individual patient and offers that "the physician is obligated to the patient and to no one else.... [T]he doctor is, as it were, alone with his patient and God" (238). Instead, we may see clinical practice as a cooperative enterprise in which patients, physicians, and a host of others are embedded in a complex system of care, interdependency, and advancing medical science. Jonas identifies an elite group that should participate as subjects in research, while the ill and the disadvantaged are exempt. Yet, today we are concerned that marginalized groups may be underrepresented in research studies, leaving them with less of the benefit of relevant scientific knowledge to guide their care.

Perhaps the problem with Jonas's vision was that his conclusions have been inappropriately applied to a much wider range of activities than what he had in mind. Jonas rejects the use of human subjects in experimentation that he sees as requiring an unreasonable sacrifice. But many of the activities that are routinely included under today's broad regulatory definition of research do not involve the kind of sacrifice that concerned him. Instead, under regulatory oversight, the reasonable burdens and benefits of research seem quite consistent with his general social contract approach. Taking his words on their face, and as a point of reference for research regulation, however, seems to have led research regulators astray.

What follows are arguments for taking an opposing stance on fair subject selection, a position more closely aligned with Emanuel et al. First we present three related principled arguments to support general participation in research, then three prudential ones.

The Argument from Justice: The musical group The New Radicals have a song, "You Only Get What You Give." The insight expressed in that title may be especially true of medicine. We all want to share in the benefits of medical science. No one can be certain about the specific nature of her own or her loved ones' future medical needs, and we would surely want to partake of the benefits if and when the need arose. Foreseeing such eventualities, we should each be expected to do our fair part by contributing to the advance of medical science that requires research with human subjects. Our bodies and biological samples, unlike money, are not always fungible. Some physical requirements are crucial for some research projects (e.g., only women, those with inflammatory bowel disease, or a serious burn can participate). So, doing our fair share may require giving of ourselves and our microbiomes in the most literal sense. Refusing to do so would be acting as a "free rider" and accepting the benefits achieved through taking advantage of the kindness, trust, and spirit of solidarity that others who do their part in advancing biomedical science contribute to the common good.

The Argument from Beneficence: If you were in need of beneficence from others, you would certainly want them to help you. Also, you can appreciate that others may at times want help from you. Specifically, they may want you to participate in research that might provide insight into their disease or the disease of someone who is dear to them. Because we accept reciprocity as the basic rule of morality, you should treat others as you would be treated. Since you would want others to help you, when the beneficence required is your participation in research, you should participate.

The Argument from a Duty to Yourself: We can all foresee ways in which we may become impaired and thereby impeded in our ability to do all of the things that we would otherwise want to do. We may have a family history of Alzheimer's disease, or Crohn's disease, or obesity, or be aware of how we might become impaired through accident or from an infectious disease. We could, however, take measures that would improve our chances of averting enduring impairments associated with those conditions. For example, we could participate in research aimed at developing improved understanding of the etiology of the disease or treatment to ameliorate the effects of an illness.

To the extent that we accept that we should take measures to keep ourselves fit and healthy so that we can complete our projects, achieve our goals, and fulfill our personal commitments, or even just to avoid becoming a burden to others, then we should take those measures. Because participating in research is a reasonable measure to prevent or lessen the impact of impairment, by extension, as a duty to oneself, one should participate in research.

The Argument from Necessity: Large numbers of diverse humans are needed as participants in biomedical research to advance medicine. To accomplish that goal, people must embrace a willingness to collaborate in the project and solidarity with the research enterprise. Without those attitudes and the actions that follow from them, studies are unlikely to meet their recruitment targets and, hence, will be unable to produce valid biomedical information for improving the practice of medicine. Subjects are needed as participants in studies to help researchers learn about the workings of the human body, its interactions with the environment and microbiome, the mechanisms of disease, the activity of drugs, and the effectiveness of interventions so that clinician-scientists can develop more effective and less burdensome treatments.

The Argument from Self-Interest: Research allows medicine to advance by providing clinicians with a better understanding of disease and by discovering interventions that allow patients to be healthy and avoid or mitigate some of the problems associated with disease and disability. We do know, however, that medicine's current understanding of the functioning of the human body, of the human microbiome, of disease, and of cures is very limited and often inadequate. Even when a good cure is available, it typically works for only a majority of cases, not for everyone. Microbiome research, for example, is intended to inform researchers about the taxa that cohabit with our genomes, how the communities of microbes interact with our genomes, and the mechanisms involved in their interaction. In sum, the studies will allow researchers to develop an understanding of which interactions are beneficial, which are not, and what can be done to transform an untoward interaction into a benign or useful one. If you and others like you do not participate in the studies, whatever is learned may not be applicable to you. Problems that may be unique to a group that includes you may not be discovered and addressed. If you do not participate in studies, you will be more likely to be left among the non-responders who derive no benefit from the interventions that are developed. In other words, you have to be in it to win it. Unless you and others like you participate in studies and contribute your samples, doctors won't know about your genes and your microbiome or about the impact of the interventions they identify on you or your kind (e.g., Caucasians, women, elderly, obese).

The Argument from Advocacy/Voice: Those who are involved in an activity have a privileged position for speaking out and demanding to be heard. Having contributed to the effort and having an insider's view of what is being done confer some measure of legitimacy and authority. Participants know something about their experience. Those who want a voice in forging the research agenda or changing research practices in some way can speak more effectively and demand attention with greater authority when they have already done their fair share. Those who are involved in the research enterprise can recognize problems in the recruitment of subjects or the actual conduct of research. They can effectively advocate for research to address particular needs because their involvement gives their voices authority. In fact, disease cohort advocacy groups do improve research and clinical practice because they are able to make themselves heard.

4. Favorable Risk–Benefit Ratio: This apparently benign principle is also controversial and complex. It involves questions about whether research should provide participants with benefits, and what can count as a benefit. A closely related issue is the acceptability of randomized placebo-controlled studies. Risk–benefit assessment also raises questions about whether inducements for research participation are ethically acceptable. Furthermore, it raises questions about whether individuals from vulnerable groups should be subjects in research and the limitations that should be placed on their participation. We shall discuss this set of interrelated issues in turn.

Benefits: The view that research is supposed to benefit participants derives from the Declaration of Helsinki: Recommendations Guiding Medical Doctors in Biomedical Research Involving Human Subjects. This document of the Word Medical Association, produced by doctors for doctors in 1964, declares in its principle III, 4, that "In research on man, the interest of science and society should *never* take precedence over considerations related to the well-being of the subject" (emphasis added). In making this claim, it ascribes an inaccurate and misleading myth to the practice of medicine by presuming that doctors should always aim at the best interest of the individual patient. To the contrary, however, exemplary practice of clinical medicine often considers the benefits of others and sometimes legitimately puts such concerns ahead of the best interest of an individual patient. Thus, some appropriate decisions made in clinical medicine do not provide benefit to every patient. Consider the patient whose surgery is delayed because another with more urgent need is allotted the operating room, or the one who is denied listing on the liver transplant list because of a low chance at long-term survival posttransplant, or the one who is quarantined to prevent him from spreading contagious disease. These patients are not benefited by appropriate medical decisions, and their best interest is not served. Thus, extending the inaccurate and misleading view of the ethics of medicine to research compounds the error. Ethical medical practice does not always require serving the patient's best interest or providing direct benefit to the individual patient, and neither does ethical research practice.

Emanuel et al. (2000) present the conditions that studies must meet to have a favorable risk–benefit ratio. Their most basic condition requires researchers to employ risk assessment standards from clinical medicine to identify possible risks to subjects, to minimize those risks, and to avoid exposing subjects to risks that are not necessary to achieving its scientific goals. Another important condition that Emanuel et al. identify involves an assessment of proportionality. Although the risks and benefits involved in studies are rarely commensurable, those contemplating enrolling human subjects in a study must examine all of the kinds of burdens involved and compare them to the array of possible anticipated benefits. Researchers must judge whether the benefits to the individual outweigh the risks and be prepared publicly to justify their conclusion.

Emanuel et al. appreciate that it is easier to justify studies when the potential harms involved are unlikely and low risk. As the risks increase, the potential benefit to subjects should increase. What Emanuel and his colleagues do not appreciate is that there may be extreme circumstances in which people have very little to lose by participating. Sometimes people with little to lose may have the possibility of gaining a great deal by participating in a study. When death seems imminent, an investigational therapy could offer a slim chance at life. In other circumstances, there may be no possibility of the patient benefiting directly, but the possibility of contributing to knowledge and the brief duration of the investigational procedure could justify enrolling subjects in the study.

Like most research ethics policymakers, however, Emanuel et al. maintain that the contribution to society is not a sufficient justification of research, but that the individual participants must also receive direct personal benefit from their participation and that only "health-related potential benefits" should count. This position, which reflects the thinking of many authors, has dramatic and counterproductive implications for biomedical science and for society. For example, the requirement that participants receive direct "potential health-related benefits" eliminates the possibility of recruiting healthy volunteers for studies because, by definition, healthy volunteers have no medical conditions that can be improved by their participation. Yet, the use of healthy volunteers is often critical in developing an understanding of normal human function, for example, in defining a healthy human microbiome. The participation of healthy volunteers is also an essential feature in assessing the safety of products, including probiotics, and sometimes a critical element in establishing an acceptable limit for risks in clinical trials.

In addition, the goal of research is to produce knowledge that can be used by all. In that sense, taking a long view, everyone who may someday come to rely upon what is learned benefits. When hypotheses are either proved or disproved by a study, we all benefit from advancing the science. When we take social benefit seriously and accord it its proper and significant weight, and when we accept that there is no ethical requirement to provide direct health benefit to research participants, then we can arrive at a reasonable and balanced view of benefits.

Randomized Placebo-Controlled Trials: A further problem of requiring research participants to be provided with "health-related potential benefits" arises in randomized placebo-controlled studies. In this gold standard for study design, all of the participants who are randomized to the placebo arm receive no "health-related potential benefits" because they do not receive the studied intervention. Yet, placebo-controlled studies are often the best model for efficiently and validly scientifically proving (or disproving) a hypothesis. By obtaining definitive findings from a study of the safety and efficacy of a new intervention with as few subjects as possible and in as short a time span as possible, precious research dollars are conserved. And when the results are positive, the intervention will be available to benefit needy patients sooner.

Individuals living with chronic illnesses such as asthma, allergies, obesity, lupus, eczema, or Crohn's disease, all conditions that may be related to their microbiome, typically anticipate decades of suffering the burdens associated with their medical conditions. Almost always, a randomized placebo-controlled trial (RCT) will provide the best evidence for the effectiveness of a newly proposed intervention (Freeman et al. 1999; Miller and Brody 2002; Temple and Ellenberg 2000). Nevertheless, IRB members and policymakers often object to RCTs, mistakenly believing that these studies will subject participants in the placebo group to undue risks and harms (Macklin 1999). In fact, the participants in the placebo group are typically subjected to the fewest risks and harms because they are not exposed to the unknown risks of the studied intervention.

To protect human subjects, those engaged in research oversight sometimes deny patients with chronic illness the opportunity to participate in optimal studies by rejecting RCT study designs. From the point of view of people with chronic conditions, it is in their long-term interests to participate in definitive studies. Yet, participants who are suffering with disease are never given a choice about whether to participate in a scientifically more or less valuable study; they are simply offered the study as approved by the IRB (Moros and Rhodes, 2010). Similarly, in the name of subject benefit, studies are sometimes ended prematurely. Subjects are never informed about how decisions for early termination of a study are made. At the same time, a well-designed and completed study that can definitively answer the study question generally would be the choice of those with a chronic condition. An incomplete study exposes participants to unjustifiable burdens and harms because it provides no knowledge. Again, designing studies with a favorable risk–benefit ratio should not support actions that expose participants to risks without allowing the study to achieve the full anticipated knowledge benefit.

Inducements: Again in keeping with other research ethics policymakers, Emanuel et al. (2000) claim that "extraneous benefits, such as payment, or adjunctive medical services... not related to the research, cannot be considered in delineating the benefits compared with the risks" (p. 2705). Yet, offering substantial financial compensation to volunteers that more than offsets the costs of transportation, childcare, and lost wages associated with study participation is one way to increase participation and improve diversity. Financial inducements are also likely to increase recruitment and retention. And financial inducements could amount to a very significant benefit in the eyes of participants.

In contrast to the challenges of recruiting adequate numbers of research participants in the United States and Europe, studies conducted in economically developing countries typically recruit adequate numbers of volunteers quickly, and the volunteers typically complete the study. Volunteers for clinical trials conducted in developing countries come from socioeconomically disadvantaged groups (c.f. Ballantyne 2008; Mfutso-Bengo et al. 2008). There, participants often volunteer for clinical trials in exchange for financial compensation or in the hope of obtaining access to medical care that might otherwise be unavailable to them. These observations have led many activists and ethicists to voice concerns that paying research participants or providing them with access to medical care may be an "undue inducement" (Macklin 1981). Bioethicists worry that monetary compensation or access to ancillary medical care might tempt individuals to accept otherwise unreasonable risks of study participation, particularly when they come from communities where a majority of people live in poverty and where the existing health care systems are fragmented (c.f. de Zoysa et al. 1998; Macklin 2004; Participants in the 2001 Conference on Ethical Aspects of Research in Developing Countries 2004). For research trials conducted in developing countries, for example, the Council for International Organizations of Medical Sciences warns that "[monetary] payments should not be so large... or the medical services so extensive as to induce prospective subjects to consent to participate in the research against their better judgment" (CIOMS 2002).

An inducement is problematic only if and when an incentive is so appealing as to impair judgment and cause an individual to ignore or discount obvious significant risks associated with study participation (Emanuel 2004; Emanuel, Xolani, and Herman 2005). Although there is scant research to support any conclusion about the moral hazard of offering inducement, a study by Halpern et al. suggests that although inducements are more likely to move poor people than well-off people to enroll in time-consuming low-risk studies, poor and well-off people are equally unmoved to enroll in high-risk studies (Halpern et al. 2004). The question of whether or not people will take on unreasonable research risks when they are paid a lot is, in the end, an empirical matter that requires further study. As a society, however, we do accept the practice of offering inducements to encourage some people to take on risks that go beyond those encountered commonly in daily life. We allow people to volunteer for military service (and receive hazard pay), to sign onto playing football, to become fishermen, and to work on oil rigs. People are even paid more for especially high-risk assignments. And when we truly value liberty, we accept that people must be allowed these options. To maintain that the relatively small inducements provided for research participants should be singled out for radical restrictions because they are more likely to distort judgment and undermine autonomy than these other acceptable inducements calls for evidence that explains why this area is different from others, or at least an argument. None has been provided.

Furthermore, in today's research environment, study proposals are reviewed by IRBs. IRBs' main task is to assure that the risks to study participants are reasonable. Their institutional reputation for conducting biomedical research according to the highest ethical standards and preventing unreasonable risks to research participants is at stake. In light of this significant safeguard, participants are already protected from having their judgment distorted by inducements and unreasonably accepting undue risks because studies that might impose unreasonable risks are not allowed.

The Vulnerable: Beyond their concern over encouraging people to participate in research with words or inducements, policymakers are especially reluctant to allow research involving "vulnerable" populations, presumably to protect them from risks and harms (U.S. Department of Health and Human Services 2011). According to the policies of the Office of Human Research Protection, and those of IRBs that adopt additional policies inspired by the regulations, vulnerable groups include the mentally ill, the mentally handicapped, pregnant women, fetuses, products of in vitro fertilization, children, prisoners, the elderly, people who are in the midst of a medical emergency, and the educationally or economically disadvantaged. That's a lot of people. Sometimes, classifying individuals as part of a "vulnerable" group that requires special protections (e.g., the elderly, the educationally or economically disadvantaged) may be inaccurate. For example, it is disrespectful to presume that an elderly person without a high school diploma cannot make decisions for himself. In any circumstance other than research, we would call it ageist and discriminatory.

Some vulnerable groups include only individuals who lack decisional capacity (e.g., young children; profoundly retarded, demented, or unconscious individuals). Individuals in such groups are vulnerable to being treated thoughtlessly and carelessly. Others need to be concerned with their interests and protect them from unreasonable harms. When it comes to research, however, it may be especially important to involve these individuals in studies. In some circumstances, it may be important to learn about their underlying debilitating condition; in other cases, it may be important to learn about how an intervention affects them in particular. And when individuals from vulnerable groups are housed together in group homes or other institutions, it may be especially important to study changes in their microbiomes to ward off disease.

Furthermore, many states, institutions, and IRBs do not permit surrogate consent for the enrollment of individuals who are unable to consent to research (Gong et al. 2010). This constraint can make the participation of people from "vulnerable" groups impossible. Yet, surrogate consent is accepted for even the most dramatic life-changing decisions (e.g., nursing home placement) and the most dangerous and drastic medical treatments (e.g., limb amputation), albeit for the patient's good. It therefore seems unreasonable that surrogate consent for enrollment in IRB-approved studies is not sufficient protection from harm for these vulnerable subjects. Without their participation, we are unable to advance biomedical science to help people like them.

Certainly, we all accept that children should be protected from avoidable risks and harms, including those associated with research. We worry that children's small and developing bodies may be especially susceptible to harms. Because children have a longer lifespan ahead of them than adult subjects would, children may have to endure any untoward effects of research longer than other subjects would and experience consequences that would not appear in those with fewer years of life remaining. These are all legitimate concerns.

Regulators have, therefore, restricted studies involving children, particularly studies that involve "more than minimal risk," that is, hazards that are greater than those encountered in the child's everyday life. Even when studies are technically allowed by current policies, the paternalistic protectionist tenor of today's research environment leaves some researchers reluctant to undertake studies involving children. Yet, children are sometimes injured and they sometimes become ill. Without study, we cannot improve the health care of children (Kopelman 2012). Studies of adults may not be applicable to children because their bodies are smaller, still growing and developing; their metabolism is different; they are far more active than adults; and their activities are different. Without data to support the treatment of children, for example, with probiotics, every treatment with an unstudied intervention is an experiment with a single subject. Thus, "treatment" with interventions that have not been studied in children exposes each child to risks without providing information that might be useful in the treatment of other future child patients. It makes no sense to expose every child to the risks of unstudied "treatment" when the alternative of enrolling a few in carefully thought out studies with vigilant oversight would expose those few to no greater risks than that of the unstudied "treatment." In effect, the current approach protects no children from risks and harms and prevents every child from receiving the benefits of scientific advance. This approach cannot represent a favorable risk–benefit ratio.

5. Independent Review: When human subject research imposes risk of harm on subjects, independent review by a panel of experts and community members assures participants and society that the study conforms to the highest ethical standards. In the United States, granting agencies, data and safety monitoring boards, and institutional, regional, and private IRBs review and oversee the conduct of human subject research. Reviewers have the responsibility to approve or withhold approval from proposed studies, to amend them and to terminate them when appropriate.

Independent review of proposed studies requires scientists to disclose what will be done to whom, how, when, where, and why. This disclosure allows reviewers to consider all aspects of a proposed study and evaluate its merits and to reach a decision about whether or not it conforms to ethical standards. Furthermore, while a study may appear very reasonable to a researcher, independent review allows another group of eyes to serve as a check on researcher bias. By giving reviewers the responsibility to assess the risk–benefit ratio of a study, independent judgment has to confirm the reasonableness of the risks and burdens involved in study participation so that unreasonable studies are not performed.

Independent review and oversight of the conduct of hazardous studies ensures the trust and trustworthiness of biomedical research by promoting confidence that nothing about the project is being concealed and that informed reasonable people find that the study design is valid, that the promised results will be valuable, that subjects are fairly selected and treated with respect, and that the risks involved have been minimized and justified and are proportional with the anticipated benefits.

6. Informed Consent: As we noted previously, informed consent has become ensconced as the cornerstone principle of research ethics[4,5] (Emanuel, Wendler, and Grady 2000). Yet, prior to World War II, few recognized the moral importance of informed consent and, consequently, it was not a common feature of research practice. In fact, very little research involved fully informing subjects about what was to be done to them and eliciting their consent to participate under those conditions.

According to bioethicists, the concept of respect for autonomy explains the centrality of informed consent in the ethical practice of research (Faden and Beauchamp 1986; Veatch 1987). According to the eighteenth-century philosopher Immanuel Kant, autonomy is the ability to take responsibility for the actions that one freely chooses. It involves a complex set of capacities: to understand the relevant facts, to appreciate their significance and how they apply, to recognize the foreseeable consequences of one's choices and their likelihood, to have relatively stable values and to make choices that reflect those values and priorities, to reach a decision, and to abide by it. Those who have *all* of these abilities can be good rulers over themselves and their choices should be respected for that reason; morality requires that we respect their autonomy. The preferences of those who do not have *all* of these abilities do not reflect autonomy. They (e.g., young children, profoundly retarded individuals, demented individuals, individuals in the grips of serious mental illness) cannot be good rulers over themselves because they lack relevant decisional capacities. Thus, they cannot be held responsible for what they do. While their preferences can be taken into account, they do not merit moral respect.

Because knowing and understanding the relevant facts is a critical element in autonomous choices, people with the capacity to rule themselves need to have critical information when they are making important choices. When researchers ask people capable of making decisions for themselves to participate in a study, researchers must provide the information that participants will need so that they can make participation decisions that reflect their own values and priorities. In effect, then, when autonomous individuals are well informed about the goals of a study and the burdens and risks involved in participation, and they consent to participation, they take on a responsibility to do what they agree to do and can be held ethically responsible for their choice. If they had been inadequately informed, either deliberately or carelessly, or even deceived about critical details of the study, their agreement to participate would not be autonomous and, ethically, they would not be held responsible for their enrollment decision.

Eliciting informed consent is, therefore, important for the ethical conduct of research, because it allows autonomous individuals to make choices in light of their values and priorities, and it shows the researchers' respect for their moral capacity. Informed consent also allows participant assessment to serve as a check on researcher bias by allowing participants to exercise their own judgment as to the reasonableness of the risks and burdens involved in study participation so that unreasonable studies are not undertaken. Informed consent is the principal mechanism for permitting people liberty and ensuring respect for participant autonomy. These reasons for valuing informed consent for research all support an important role for informed consent in the ethical conduct of human subject research. They explain its importance in light of other important features of the research enterprise: maintaining public trust in science and research through honest and transparent practice, ensuring that avoidable harms and burdens are not imposed on participants, and treating others with the respect and dignity that they deserve.

In sum, informed consent is an important consideration in the ethical design of human subject research. It is not, however, "absolutely essential" as the Nuremberg Code and other documents have since maintained. In fact, in some circumstances there may be strong overriding reasons for proceeding with research without informed consent. Often, surrogate consent may be sufficient for enrolling a subject in a study. Sometimes, for example, in emergency situations, studies may ethically proceed without any consent. Under the "Final Rule" of current U.S. policy, emergency research that meets strict guidelines is now allowed to proceed without informed consent. In such circumstances we rely on a well-accepted concept from political philosophy, namely, presumed prior consent. When a choice is eminently reasonable, the kind of choice that no reasonable person could refuse to endorse, we presume that the individual involved is such a reasonable person, and proceed *as if* the individual involved had actually made the choice (Scanlon 1998). Research that promises a chance for significant benefit to the individual or society, or studies that involve risks that are miniscule or harms that are only short lived, can be justified with the concept of hypothetical presumed prior consent.

*7. **Respect for Enrolled Subjects:*** People who enroll in biomedical research often accept some burdens and some risks. They are expressing their willingness to promote the social good of producing biomedical knowledge to be shared with others. Regardless of the participants' actual motives, participation in research should always be seen as a noble choice that expresses concern and sympathy for the plight of others and solidarity with the community of humanity. People who volunteer to serve as research participants should not be seen as merely subjects being used for the purposes of others. Instead, they should be acknowledged as collaborators in important social projects, as courageous citizens who accept their responsibility and do what they can to further biomedical research. Research participants are entitled to feel proud of doing their part, and that they have earned respect from others for their sacrifice.

Research participants should, obviously, be treated very well. They deserve careful treatment and monitoring during the course of a study to ensure that risks are minimized and that their well-being is maintained. Also, because the burdens of a study may turn out to be more significant than a subject had anticipated, thereby changing the risk–benefit ratio, subjects must be allowed to withdraw from a study without penalty. Beyond that, as Emanuel et al. (2000) note, information that they disclose to further the research project should be safeguarded according to standards for medical confidentiality (p. 2707). And because research participants are collaborators in the research enterprise, they deserve to be informed of study findings so that they can partake in the pleasure of having helped to produce valuable knowledge. Furthermore, excellent treatment of research subjects throughout a study helps to promote and maintain society's trust in the research enterprise.

Privacy and Confidentiality: As a feature of showing respect, we often believe that we should show respect for people's privacy.[6] This concern is directed toward research participants and, therefore, requires some clarification. Privacy is a concept in common morality. It identifies areas that are safeguarded from the scrutiny and intervention of others, areas that are marked off from public space by natural boundaries. Private domains typically have visible boundaries. Whatever is in and on my body, in my head, in my room, in my home, and in my mail are all private. Private domains are protected from the intrusion of others by laws, social practices, and social sanctions. In common morality, however, privacy protection is not absolute and not guaranteed. Most importantly, private domains are protected from unwanted government intrusion, although we allow exceptions from privacy protection for the sake of the public good (e.g., we allow the police to conduct search and seizure operations). Aside from protection from government intrusion, the task of maintaining privacy is largely left in the hands of individuals who may or may not share their thoughts, conceal their bodies, close their blinds, or leave their opened mail in places where others can easily read it.

In recent years we have seen privacy protections (i.e., the Health Insurance Portability and Accountability Act of 1996 [HIPAA]) enacted. These measures are designed to guarantee that there will be no breaches in privacy. Some HIPAA requirements are unreasonable in that they fail to take into account other reasonable and legitimate social and personal goals, such as advancing the social good and protecting others from harm. They are also paternalistic in that they impose protections for the good of others that many may not want. For example, I may not care about protecting the privacy of my biological samples; I may care far more about advancing biomedical knowledge than I do about the absolute protection of my privacy; and I may care most about helping my fellow man.

The concept of "privacy" should be distinguished from the concept of medical "confidentiality." Whereas privacy is a common morality concept, confidentiality is a concept from the professions: the priesthood, law, and medicine. Society relies on the ethics of these professions to uphold confidentiality. The concept of confidentiality defines a protected space for professional interactions where privacy is safeguarded from the scrutiny and intervention of others. That space is marked off from public space by artificially constructed boundaries, and the rules for sharing confidential information are defined and enforced by the professions.

---

[6] The concept of privacy is discussed at length in chapter 4 of this volume.

In medical treatment, medical education, and biomedical research, information about people should be treated according to standards of confidentiality that govern other medical interactions. Information is to be shared on a need-to-know basis. We maintain that confidentiality, rather than privacy, should be the standard for biomedical research. Confidentiality is already the prevailing standard in research practices that are given other names: public health surveillance, quality assurance, quality improvement, and medical registries. In none of these long-standing practices has sharing information according to standards of medical confidentiality caused harm. These studies are all conducted with professional oversight, institutional oversight, and in accordance with professional ethics. Furthermore, in this age of new research technology, genomics, personalized medicine, and human microbiome research, the need for the involvement of a broad spectrum of the population makes it especially important to remove barriers to research and promote broad research participation.

Protectors of privacy are especially concerned about protecting individuals who have donated to biobanks and those whose biological samples, left over from other studies or from clinical care, can be used additionally in research.[7] Samples linked to health records or other identifiable information are far more valuable than anonymized samples that have been stripped of any possible information that could possibly identify the donor source. Anonymizing data entails losing a significant portion of the sample's research value because it prevents investigators from checking their hypotheses against the donors' clinical conditions or recontacting donors to learn about their evolving health status. Furthermore, completely anonymizing data is not always completely possible.

To circumvent some privacy restrictions and research regulations, some people who operate biobanks and sample banks have also resorted to describing biobank samples as nonhuman subject research. This redescription obfuscates what is actually being done and fails to address the real problem, the unreasonable concern for protecting privacy. Privacy protectors also fail to recognize that we already have a useful and appropriate standard of confidentiality that provides effective and adequate protection of people's personal information. When medical confidentiality is upheld in medical treatment and research, people are not harmed. At the same time, research conducted within the constraints of confidentiality provides a significant public good.

The goals of promoting patient, participant, and public safety, as well as providing for the public good, may at times be more important than preserving privacy. When we keep the research benefits that we desire in mind and uphold appropriate confidentiality limitations, data from biobanks and sample

---

[7]The issue of protecting personal information in biobank studies is discussed at length in chapter 6 of this volume.

banks may be shared so as to significantly increase the research use of samples. Identifying information can be limited to reflect the need to know (including the need to recontact). Where possible, and with informed consent when samples are taken, materials remaining from clinical uses and other research uses should be available for additional research purposes. For the most part, the use and reuse of biological samples involves only *de minimis* risk, that is, hard-to-imagine and highly unlikely harms, primarily associated with a remotely possible loss of privacy.

Implications for the Conduct of Human Subject Microbiome Research

1 and 2. Value and Validity: All of the conditions discussed previously for conducting human subject research according to high ethical standards are relevant to research on the human microbiome. Collections for biobanks and sample banks are intended to provide research materials for future studies. The more these repositories are used, the better, in terms of the increased possibilities for producing more knowledge, and more valuable findings. Hence, policies governing the use of biobanks and sample banks should actively promote their use in order to advance science and minimize obstacles for accessing samples. Policies should promote sharing of samples and data, investigational techniques, and research findings.

Similarly, studies that aim at elucidating the connection between the human microbiome and disease resistance might improve clinicians' understanding of inflammation processes, disease, and disease prevention. And interventional trials of prebiotics (interventions designed to encourage the growth of certain organisms) and probiotics could contribute significantly to the armamentarium used to promote health and improve people's quality of life. Also, studies of phages and lytic enzymes could produce important tools for combating infectious disease. These sorts of studies related to the human microbiome must be conducted according to rigorous principles of scientific design so that their results will be valid and valuable.

3. Fair Subject Selection: Studies to map the myriad interactions between human beings and the commensal, symbiotic, and pathogenic microorganisms that inhabit the body, like the HMP, are likely to improve our understanding of how the microbiome helps to maintain normal human physiology (Turnbaugh et al. 2007). Such projects may also identify how specific changes in these interactions can contribute to human pathology and lead to the development of new markers to predict incipient disease, new interventions to prevent disease, and new drugs to treat disease (Kinross et al. 2008; Wilson 2009).

Achieving these lofty research goals, however, will require complex retrospective, cross-sectional, and prospective studies designed to characterize the microbiome within the human population. Researchers will need to collect microbiological specimens from a wide spectrum of volunteers to address such key questions as: Do all humans share a core microbiome? How is this core microbiome affected by genetic differences within and between different human populations? How is the microbiome affected by differential environmental exposures? And how does the microbiome evolve over time with changes in age, diet, and lifestyle (Fierer et al. 2008; Grice et al. 2008; Turnbaugh et al. 2008; Zaura et al. 2009)?

A significant challenge for microbiome research will be ensuring that specimens are collected from a sufficiently diverse number of study participants, particularly with respect to age, gender, race and ethnicity, health, socioeconomic status, and environmental exposures. Determining with relative accuracy the nature of the core microbiome, for example, will require measuring the types of microorganisms present in various parts of the human body (including skin, nasal and oral cavities, gastrointestinal tract, and male and female genital tract) from a well-defined sample of demographically, geographically, and culturally diverse individuals. Detecting changes in the microbiome that are associated with the onset or progression of acute or chronic diseases will be even more difficult and require association studies among individuals with particular illnesses, again from a demographically, geographically, and culturally diverse pool of study participants. Ensuring diversity among study participants, thus, is necessary to ensure that the results of these microbiome studies are generalizable to large segments of the human population. Inadvertent or systematic underrepresentation of demographically, culturally, or geographically defined groups is likely to affect the validity and broad applicability of these and future studies. Research designed to characterize the nature of the core human microbiome may miscalculate microbiological diversity when some segments of the population are under- or overrepresented. Similarly, association studies may miss key changes in the human microbiome associated with disease onset or progression. Such errors are likely to occur in current association studies that fail to enroll sufficient numbers of at-risk individuals from particular racial, ethnic, or socioeconomic groups. In turn, relying on erroneous characterizations of the core microbiome in future studies can compound the errors.

Unfortunately, important ethnic, demographic, and socioeconomic groups are routinely underrepresented in most research studies. A recent study of American participants in cancer-related clinical trials, for instance, found that racial and ethnic minorities, the elderly, rural residents, and individuals of lower socioeconomic status were underrepresented (Ford et al. 2008). A similar analysis of epidemiological studies in Europe found that nonparticipants were likely to be poor, to be members of ethnic minority groups, and to have physical or mental health issues (Drivsholm et al. 2006). There have also been a number of surveys that aimed at identifying barriers to study participation within various underrepresented groups, including a large NIH-funded survey of 6,000 people with cancer who agreed or declined to participate in clinical trials (NCI 2002). Among members of racial and ethnic minorities, many were reluctant to participate in research studies because of lingering distrust of the medical and research establishment in light of past abuses (e.g., the Tuskegee Syphilis Study). Older individuals, those with lower incomes, and those who live in rural communities also have difficulty in getting to clinical trial sites, either because of transportation and childcare constraints or because they cannot afford to take time off from work.

General participation is needed in medical registries to record outcomes for all patients using new prebiotics, probiotics, and phages, and using them in new ways. Probiotic registries would allow clinicians and manufacturers to follow products and learn about problems associated with their use (e.g., that the drug that corrects one problem causes another, that the product stops working after a period of time). Without collecting data on all users, it takes longer to identify such problems, particularly when they affect only a subgroup of users. With data on complications, unforeseen problems can be identified and addressed in a timely way.

Currently researchers are working to develop personalized medicine, that is, ways to use information about an individual's genetic makeup and microbiome and use it to select the specific treatments that are most likely to be effective in treating that individual's medical problems. This technology will allow doctors to match the right treatment to the patient and avoid the wasted time, money, and risks involved in administering a treatment that works well in most people but not in some individuals. The technology cannot be developed, however, without first collecting and analyzing a tremendous amount of genetic data. Again, broad participation will be necessary. Biobanking and sample banking, efforts to collect data on the genomes and microbiomes of large populations, will be critical in advancing personalized medicine because only with large samples will scientists be able to identify the distinguishing differences. Information on large populations must be collected and analyzed to inform scientists' understanding of variations in the human genome and microbiome that account for differences in disease resistance and susceptibility and treatment efficacy.

Thus, microbiome and genetic research, far more than previous clinical trials, require broad participation. If we want to reap the benefits that this new science can provide, we will need many people from a broad spectrum of the population to participate. So, in addition to researchers making a sincere and sustained effort to enroll a broad spectrum of subjects in microbiome studies, we also have to consider how the general population should regard participation in microbiome studies. The arguments presented in the previous section demonstrate that there are good reasons to participate in biomedical research and perhaps even a moral duty to participate. In this light, and without going so far as to make participation mandatory, our society should acknowledge the moral duty and shift our moral perspective from the Jonas view, that asking people to participate in research is exploitative conscription, to seeing the request as asking people to do what they should. Instead of seeing research participation as entirely optional, it should be seen as doing one's fair share. This change in perspective would encourage broad participation in scientific activities to map the human microbiome and elucidate its role in maintaining health. It would promote research to ascertain the efficacy of probiotics, prebiotics, phages, and lytic enzymes and give researchers support and confidence in recruiting people with and without disease.

Some of the people who will be encouraged to enroll in studies of the human microbiome will come from vulnerable groups. In some cases, it may be especially important to study these individuals because there may be specific microbes shared among individuals who live in group facilities (e.g., nurseries, hospitals, nursing homes, prisons) or because changes in the microbiomes of those living in group facilities are more likely to have an impact on cohabitants. Furthermore, the institutional environment may have a unique impact on its residents' microbiomes, or the treatment for some (e.g., with an antibiotic) may have an impact on the microbiomes in other cohabitants. Mechanisms to both protect these vulnerable individuals and permit the conduct of reasonable studies have to be established.

4. Favorable Risk–Benefit Ratio: We know that most of the sample collection involved in microbiome research is likely to involve very miniscule physical risks. In most cases, the risks are so small, so little even compared with the risks of everyday life, that we call them *de minimis*. Although some authors have cautioned about risks to privacy in biobank research, with thoughtful policies in place and oversight, the likelihood of breaches in confidentiality is very small. The national Genetic Information Nondiscrimination Act (GINA) rules reduce the risk even further. We therefore recommend extending something similar to GINA protections to all biobank and sample bank research. So long as the risks involved in these studies are *de minimis* and the value and scientific validity of the studies are maintained, the benefits in biomedical knowledge make the risk–benefit ratio of microbiome research overwhelmingly in favor of moving forward.

Some study-related procedures involved in microbiome research are likely to be uncomfortable or embarrassing (i.e., providing stool specimens or genital tract swabs). That could make people reluctant to participate. Yet, we also know that financial inducement can be a powerful motivating agent and can overcome reluctance. For microbiome sample bank research, concerns about financial compensation and undue inducement are, for the most part, misplaced. Although different people will have different reasons for volunteering for these studies, ranging from altruism and curiosity to financial need or a desire for access to quality care, these motivations are neither unreasonable nor ethically suspect. Even a relatively strong incentive to participate in a microbiome study, such as (perhaps) $500 in financial compensation for a single stool specimen or vaginal swab, is not inherently problematic because the risks involved are *de minimis*. Although study participants may be inconvenienced or embarrassed, they are unlikely to experience any enduring study-related risks beyond, at most, those associated with a routine physical exam.

A more likely concern for microbiome mapping studies will be whether the compensation offered to volunteers is adequate to ensure that all of the different ethnic, cultural, and socioeconomic groups are represented in the study population. Given the nature of microbiome sample bank research and the types of specimens collected, it may be necessary to offer substantial financial inducements to enroll a sufficient number of participants. The socioeconomic representativeness of the participant pool, however, may vary depending on the level of compensation offered. For some studies, it may be necessary to offer significantly more financial compensation to overcome logistical barriers to participation by socioeconomically disadvantaged participants. In other studies, it may be necessary to offer larger financial incentives to entice middle- and upper-class residents to participate. In South Africa, for example, the Medicines Control Council requires that human immunodeficiency virus (HIV) prevention trials offer study participants R150 (approximately $14 US) per study visit to compensate for their time and lost wages. In the more socioeconomically disadvantaged regions of the country, this requirement is usually seen as appropriate and research sites in those communities have had little difficulty in participant recruitment or retention. In wealthier, urban regions of South Africa, however, many HIV prevention researchers have reported difficulty in recruiting study participants. In those communities, many research participants complain about being paid too little for their time and efforts. Microbiome researchers will be challenged to determine a level of financial compensation or other incentive that will ensure adequate rates of recruitment and retention of a diverse pool of study participants.

Whereas participation in biobank and sample bank research typically involve only *de minimis* risks, inducements may be more of an issue in microbiome research related to elucidating the role of microbiota in disease resistance and susceptibility and in the development of probiotics and phages. Many of these studies will involve repeated research encounters; some may be invasive, and some will involve some measure of risk. Inducements are likely to be useful in encouraging enrollment and retention. Retention will be especially important in these studies because researchers will need to see the same participants repeatedly to generate useful data. After the initial enthusiasm and goodwill have lost their edge, inducements encourage people to return to complete the study. In many cases, until a participant completes a study, no useful data are generated.

As we explained, out of concern for undue inducement, today's policies limit reimbursement for all research participants. As we see it, compliance with these policies does more harm than good. Almost all of the healthy volunteers who participate in studies do so, at least in part, for the money. In most circumstances to speed recruitment, these subjects would be paid substantially more money by researchers and the contract research organizations conducting the tests, if allowed. These research participants would all be acting voluntarily, and thereby accepting the research-related inconvenience as well as some low level of risk, all with IRB approval and oversight. Under current restrictions they are, however, being denied the benefit of the extra payment that would be forthcoming but for the concern over "undue inducements." From the point of view of research participants, if they knew that their payment was being limited by those who had to approve the studies, they would see themselves as being harmed by the protectionist thinking that guides these regulations.

In addition, theoretical and safety problems are generated by restricting payments to research participants in chronic disease studies. In many cases of chronic diseases that could be targets for probiotic or prebiotic treatments (e.g., inflammatory bowel disease), optimal study design would begin by first observing each participant for a period of time, thereby creating a reliable record of baseline function. It would be followed by slow titration of the study agent while having the subject visit a physician frequently for observation. Without compensation, however, most subjects want to get to an anticipated therapeutic dose as quickly as possible, and they do not want to be bothered by frequent trips to a study site. Thus, study design is adjusted to create adequate compliance by sacrificing both science and safety. If reimbursed adequately for their effort, however, many participants would happily permit a study to drag on. In an elongated study, the subjects would receive increasingly larger compensation for their participation and, perhaps, a bonus for completing the study. IRB reviewers who rely on current standards may worry that such an inducement would be "coercive," that is, involve an unjust threat of harm. But certainly no subject would be unjustly threatened with harm by an offer of increased compensation; participants would simply be rewarded with greater inducements for their contribution to the completion of the study. To the extent that the risks and harms involved are minimized and in keeping with a favorable risk–benefit ratio, there should be no ethical concern about providing inducements, even graduated inducements, to encourage study completion.

5. ***Independent Review:*** Independent scientific and ethical review by a funding agency, a sponsoring institution, and/or an IRB will also be important for microbiome studies. As we mentioned earlier, independent review and oversight of clinical research is critically important. It provides another perspective in addition to the researchers' on the scientific validity of the study design. Independent review also attests to the reasonableness of the participants' exposure to risks and burdens. Some microbiome studies—for example, trials of probiotics to alleviate flares of irritable bowel syndrome—will involve

some measure of risk. Vigilant IRB protocol review and oversight should ensure that risks are minimized and that no study that exposes participants to unreasonable risks is allowed to be undertaken. In doing so, independent review assures the public that the researchers are trustworthy, that their use of society's resources provides value, and that the scientists' findings can be trusted. It also gives participants confidence that nothing about the project is being concealed and that they will be treated with care and respect.

Given that background assurance, it is reasonable to expect that the worry over inducements encouraging participants to take unreasonable risks, or to continue with an unreasonably burdensome study, will be minimal. The importance of completing studies and advancing science should be given the weight it deserves and balanced against unsubstantiated presumptions about people's decision-making abilities being disabled by research incentives.

6 and 7. Informed Consent and Respect for Enrolled Subjects: Eliciting informed consent is important for the ethical conduct of research, including microbiome research. It allows autonomous individuals to make choices about participation freely and in light of their values and priorities. When researchers provide potential participants with information about the aims of the study and the anticipated burdens, they show respect for the participants' moral capacity. The information allows participants to assess how joining the study would fit with their lives and other commitments so that they may judge whether or not to enroll.

In clinical trials of probiotics and phages, informed consent is just as important as it is in other clinical studies. There may, however, be circumstances where either eliciting informed consent in microbiome research is not possible or the risk–benefit ratio of implementing it is impractical. When potential participants lack decisional capacity, surrogate consent should be sufficient for enrollment in IRB-approved microbiome studies. When eliciting informed consent is impossible (e.g., due to urgency), community consultation along with a heightened level of scrutiny (e.g., full committee IRB review) would be in order.

Because some human microbiome research requires the participation of large numbers of participants and the collection of vast amounts of data, it will be especially important to explain to the public, as well as individual study participants, why their participation is needed. The public needs to understand the value of this research so they can come to acknowledge its importance and share in the goal of advancing studies of the human microbiome. This makes public education an important component in promoting the HMP. It shows respect for the moral and intellectual capacity of those beyond the scientific community, and it also serves to further the research agenda in advancing the common good.

In microbiome biobanks, the goal is to collect samples from a broad spectrum of the population for use in future studies. The specific aims of future studies are not known at the time when samples are taken, but studies using biobanked samples involve only *de minimis* risks. It will be impractical to elicit informed consent for each new study that is undertaken with previously collected samples. Instead, at the time when a participant enrolls in the biobank and samples are collected, general information about the goals of the project, the process for approving future biobank studies, policies for recontacting participants and returning incidental findings, and assurance of confidentiality should be provided. This information falls short of standards for informed consent, but it does reflect a standard of reasonableness and show respect for participants. These issues of consent for population research and biobank research will be discussed in greater detail in the two final chapters of this volume.

Research and Regulation of Probiotics and Phages

Although research on probiotics and phages will be important in advancing biomedical science, the governing regulations present challenging economic, practical, and ethical issues. Foods containing bacteria have been intentionally consumed by humans for their health benefits for hundreds if not thousands of years. In today's world, however, marketing these products as biotherapeutics (i.e., drugs) or as probiotics (i.e., food) has to conform to a complex regulatory framework. As we mentioned earlier in this chapter, the Food and Agriculture Organization of the United Nations and World Health Organization Working Group (2002) defines probiotics as "live microorganisms which when administered in adequate amounts confer a health benefit on the host." Prebiotics are a related group of products. A prebiotic is "a non-digestible food ingredient that beneficially affects the host by selectively stimulating the growth and/or activity of one or a limited number of bacteria in the colon" (Gibson and Roberfroid 1995, 1405). In the United States, however, there is no official legal or regulatory standardized definition of a probiotic or prebiotic. Furthermore, the categories of the current FDA regulatory framework do not easily accommodate probiotics, prebiotics, or phages.

In the United States, the FDA is the organization responsible for regulating food, drugs, and a host of other products that have an effect on human and animal health. The operations of the FDA are governed by several pieces of legislation, most significantly the Federal Food, Drug and Cosmetic Act. In part, the regulations that apply to a product are related to the kinds of claims that a manufacturer makes about its contents, intended use, and efficacy. Those claims determine which of nine regulatory categories apply: drug, food, dietary supplement, food additive, biological product, medical food, functional food, device, or cosmetic. Currently, probiotics are being marketed as foods, food additives, dietary supplements, biological products, and cosmetics.

When a manufacturer claims that a product can cure, diagnose, or treat disease, or when it is "intended to affect the structure or function of the body of man or other animals" (Lee and Salminen 2009, 117), it is regulated as a drug. Before new drugs or new components in drugs can be marketed or prescribed, they must be submitted to the FDA with a new drug application (NDA). Because drugs are substances that are presumed not to be safe for human consumption, strict regulations govern the approval of drugs so as to protect the general public from harm once the drug is marketed and to protect research subjects from harm during the testing and certification process. The evaluation of a drug enables clinicians who prescribe the product and patients who use it to know how much of the drug to use and when the benefits justify the risks. An NDA involves extensive clinical trials to determine the safety of the drug, how it works, and its efficacy for the intended use. After a manufacturer completes this extensive testing, the FDA reviews the application. If approved, the drug may be marketed as safe and effective for its intended use when taken as prescribed. This process takes years.

As of this writing, the FDA has not approved any probiotics to be marketed as drugs (Lee and Salminen 2009, 119). One serious problem with conducting research on the efficacy of probiotics is related to the FDA definitional structure. Probiotics are, by definition, safe. This allows them to be sold as a food, dietary supplement, or food additive. Hence, they cannot be drugs, because drugs are unsafe. Nevertheless, claiming that a probiotic is effective in alleviating a symptom or disease is making a drug claim. That requires the product to be evaluated as a drug. For example, *Lactobacillus* GG is currently approved as a food additive that is safe for use in infant formulas (Lee and Salminen 2009, 469). To assess its efficacy in addressing symptoms, however, the product must go through all of the same multiple levels of testing and approvals that would apply to introducing a new chemotherapy as a drug. Thus, the presumption that drugs are dangerous while foods are not, coupled with the extensive review and research process required by an FDA NDA, creates an environment in which manufacturers are reluctant to gather evidence on the efficacy of their probiotic products. Instead, they market their products as foods or dietary supplements and forgo learning whether or not they are effective (Lee and Salminen 2009, 120). So long as manufacturers find a market for their probiotics without proving their efficacy, they avoid making clear claims about their function.

Similar problems are likely to affect the development of phage and lytic therapies. It is difficult to meet the regulatory requirements for approval as a drug. Also, manufacturers who are able to market their products as food additives or biologicals have little incentive to generate the evidence to prove their efficacy. As of this writing, U.S. companies have focused primarily on the development of phage products for decontamination of food, plants, fields, and livestock. In 2006, however, the FDA approved the first cocktail of six individually purified phages as a food additive when used as a treatment for

Listeria monocytogenes contamination of ready-to-eat meat and poultry products (Fischetti, Nelson, and Schuch 2006). One problem for companies trying to get FDA approval of their products is that a combination of phages may be needed to kill a single organism. Phages also pick up genetic material from bacteria. This will make it difficult to meet FDA requirements for accurately and specifically describing the product (Borrell 2012).

Problems with applying the current regulatory framework to probiotics and phages call out for the creation of a new regulatory category or mechanism. Two options are being discussed with respect to probiotics. One approach would use the existing regulatory mechanism of an over-the-counter (OTC) drug monograph. The other suggestion is to develop an entirely new standard for certifying the safety and efficacy of these new products. OTC drugs include, for example, antiperspirants, toothpastes, acne remedies, baldness remedies, condoms, and sunscreens. OTC drugs can be regulated along two different pathways, through a new drug application or through an OTC monograph. In general, OTC drugs must meet the same safety and efficacy standards as prescription drugs (FDA 2010). What distinguishes OTC drugs from prescription drugs is that consumers must be able to "self-diagnose, self-treat and self-manage" their condition, their benefits must outweigh their risks, "the potential for misuse and abuse is low," and "they can be adequately labeled"(FDA 2010). Many OTC drugs use the OTC monograph system to obtain FDA approval.

According to the regulations, foods are defined as "(1) articles used for food or drink for man or other animals; (2) chewing gum; and (3) articles used for components of any such article" (FDA, n.d.). They are regulated by Current Good Manufacturing Practices, which provide the "minimum sanitary and processing requirements for producing safe and wholesome food" (FDA 2004). Dietary supplements are subject to few FDA regulations. They are regulated under a different piece of legislation, the Dietary Supplement Health and Education Act, and considered as a special type of food. A dietary supplement is defined as "a product taken by mouth that contains a 'dietary ingredient' intended to supplement the diet. Dietary ingredients may include: vitamins, minerals, herbs or other botanicals, amino acids, and substances such as enzymes, organ tissues, glandulars, and metabolites. Dietary supplements can also be extracts or concentrates, and may be found in many forms such as tablets, capsules, soft gels, gel caps, liquids, or powders" (FDA 2009). Dietary supplements do not require any premarket FDA approval unless they contain a "new dietary ingredient" (NDI), which must be tested for safety before it is marketed. This means that only the manufacturer of a dietary supplement, and no oversight agency, ensures that its product is safe before it is marketed. The FDA takes legal action against a manufacturer when a supplement already on the market is found to be unsafe. For the most part, dietary supplements are not required to be registered with the FDA or meet Generally Recognized as Safe (GRAS) standards. Dietary supplements are, however, subject to a set of regulations called the "Current Good Manufacturing Practices" that are supposed to guarantee their "identity, purity, quality, strength and composition (FDA 2004)." Part of this standard requires that the dietary supplement not be "adulterated," meaning that it does not "present a significant or unreasonable risk of illness or injury when used under its ordinary conditions of use" (Lee and Salminen 2009, 114). Also, according to Federal Trade Commission regulations, the product label information must be truthful and not misleading.

Stricter regulations apply to dietary supplements that contain an NDI, "an ingredient not marketed in the United States before . . . the Dietary Supplement Health and Education Act (DSHEA) was passed" (115) in 1994. This legislation applies to probiotics. The manufacturer is responsible for determining whether or not its product's ingredients include NDIs. An NDI "may only be used if a) there is a history of use of the ingredient, or other evidence establishing that the proposed use of the ingredient 'will reasonably be expected to be safe,' and b) the manufacturer files a premarket notification containing the information in support of safety at least 75 days before marketing the product."

The types of claims that manufactures make about both foods and dietary supplements are strictly regulated. The FDA recognizes and regulates at least seven different types of claims for a food or dietary supplement: (1) disease claims, (2) content claims, (3) benefit or efficacy claims, (4) health claims, (5) qualified health claims, (6) structure or function claims, and (7) publications as claims. (Lee and Salminen 2009, 117–118).

1. A disease claim is "a claim to diagnose, cure, mitigate, treat, or prevent disease." Such a claim can only be made about a drug or a biological product (FDA 2002).

2. A product's label or manufacturer makes a content claim when it states that the product contains a substance or an amount of the substance (e.g., a certain bacterial strain). The FDA allows both foods and dietary supplements to make these claims. In practice, however, these claims about probiotics are often inaccurate because there is not a uniform standard for identifying the strain of a microbe and the shelf life of strains is not well established (Lee and Salminen 2009, 116).

3. Benefit or efficacy claims about foods and dietary supplements must be "demonstrated by reliable information (Lee and Salminen 2009, 116)." It is up to the manufacturer to substantiate its claims, and the FDA has the authority to enforce this rule.

4. Health claims "expressly or impliedly characterize the relationship of a substance to a disease or health-related condition" (Lee and Salminen 2009, 117). A statement such as "Diets high in calcium may reduce the risk of osteoporosis" is a health claim (Lee and Salminen 2009 117; FDA 2003). For a manufacturer to legally make a health claim about its product, it must petition the FDA and be approved for the claim. The claim has to meet the standard of "significant scientific agreement" (Lee and Salminen 2009, 117). Currently no product containing a probiotic has yet met this standard.

5. Qualified health claims (QHCs) are subject to an even more permissive FDA approval process. In 2003 the *Consumer Health Information for Better Nutrition Initiative* established the QHC category to allow manufacturers to make claims about products that fail to meet the standard of "significant scientific agreement." A manufacturer that wishes to legally assert a QHC must only submit a premarket petition and be granted FDA approval for it. The standard the QHC must meet is that "there is emerging evidence for a relationship between a food, food component, or dietary supplement and reduced risk of a disease or health-related condition" (FDA 2003).

6. Structure or function claims are extremely prevalent in the marketing of probiotics as foods and dietary supplements. These claims are often the subject of FDA enforcement when they seem to fit into the category of drug claims, health claims, or other regulated claims. Structure or function claims must always be accompanied by the disclaimer that the product has not been "evaluated or approved" by the FDA and that the product "is not intended to diagnose, treat, cure or prevent any disease" (FDA 2009.). The "not intended" clause is meant to distinguish drugs from other products.

Several different types of claims are included under this standard. The claim may "describe the role of a nutrient or dietary ingredient intended to affect normal structure or function in humans, for example, 'calcium builds strong bones'" (FDA 2003). The claim "may characterize the means by which a nutrient or dietary ingredient acts to maintain a structure or function, for example, 'fiber maintains bowel regularity'" (FDA 2003). The claim may also "describe general well-being from consumption of a nutrient or dietary ingredient" (FDA 2003). And the claim may "describe a benefit related to a nutrient deficiency disease (like vitamin C and scurvy), as long as the statement also tells how widespread such a disease is in the United States" (FDA 2003). Manufacturers must "notify the FDA of the claim within 30 days after marketing a dietary supplement accompanied with the claim." As with other claims, despite the fact that they do not require premarket approval, the manufacturer "is responsible for ensuring the accuracy and truthfulness of these claims" (FDA 2003).

One popular dietary supplement probiotic product is Align, manufactured by Procter and Gamble. It contains *Bifidobacterium infantis* and the manufacturer makes a structure/function claim about it. Proctor and Gamble advertise that Align can "Build and maintain a healthy digestive system" and "Restore your natural digestive balance and protect against occasional digestive upsets" (Align 2013). This claim may be permissible under FDA regulations. A probiotic dietary supplement or food claim 'to contain bacteria that prevent the occurrence of urinary tract infections' would, however, be making a drug claim, thereby being illegally marketed. Most people find it difficult to distinguish a structure/function claim from a drug claim. Few people know that these kinds of claims are supposed to reflect different standards of evidence, and few can discern the difference in the claims.

7. Manufacturers of dietary supplements, but not foods, may use publications as claims and distribute scientific literature as part of their marketing (Lee and Salminen 2009, 118). These publications must meet four standards: they must not be "false or misleading," they must not "promote a particular brand," they must "present a balanced view" of the scientific data, and, finally, they article must "be reprinted in its entirety" (Lee and Salminen 2009, 118).

Food Additives: A food additive is defined as "any substance the intended use of which results or can reasonably be expected to result, directly or indirectly, in its becoming a component or otherwise affecting the characteristic of food" (Lee and Salminen 2009, 112). Food additives can be regulated in two ways. One route is for the manufacturer to submit an application for premarket FDA approval. This approval process requires the manufacturer to produce scientific evidence that shows with "reasonable certainty" that the additive will not cause harm when used as directed (Lee and Salminen 2009, 112). The alternative route is to demonstrate that the additive meets the GRAS standard. By that standard the ingredient must be "generally recognized, among qualified experts, as having been adequately shown to be safe under the conditions of its intended use..." (FDA 2011a). The addition of *Lactobacillus bulgaricus* to yogurt meets this standard because this bacterium has been part of the process of yogurt manufacturing for a long time without ill health effects on its consumers. As of this writing, two probiotics have met the GRAS standard: *Bifidobacterium lactis* as an additive to infant formula and *Lactobacillus acidophilus* and *Lactobacillus lactis* "for use in fresh meat for the control of pathogenic bacteria" (Lee and Salminen 2009, 114). The FDA and U.S. Department of Agriculture have also approved several bacteriophage products to target and kill *Listeria monocytogenes* bacteria on foods, giving them GRAS status (FDA/CFSAN 2007).

Biologic Product: The Public Health Service Act applies to a wide range of biological products: vaccines, blood and blood components, allergenics, somatic cells, gene therapy, tissues, and recombinant therapeutic proteins. These products are categorized as biologic products when they are "applicable to the prevention, treatment, or cure of a disease or condition or injuries to man" (Lee and Salminen 2009, 112). It could be appropriate to include probiotics and phages in this category because they are microorganisms, hence, biologic products. Yet, according to the *Handbook of Probiotics and Prebiotics*, no probiotics have yet been "commercialized or evaluated under this category" (Lee and Salminen 2009, 112).

Medical Food: A medical food is defined by the Orphan Drug Act as "a food which is formulated to be consumed or administered enterally under the supervision of a physician and which is intended for the specific dietary management of a disease or condition for which distinctive nutritional requirements, based on recognized scientific principles, are established by medical evaluation" (FDA 2007). What makes medical foods distinct from both foods and drugs or supplements is that they are designed to meet the special nutritional needs of specific conditions rather than to treat disease or merely to feed.

Medical foods are regulated according to the same standards as foods and food additives. They are not subject to specific premarket approval by the FDA.

The probiotic VSL#3 is the most popular probiotic recommended by gastroenterologists and colorectal surgeons for the management of select gastrointestinal disorders (Cordina 2011). VSL#3 is a patented mixture that contains eight different strains of bacteria. It is designated as a medical food by the FDA for the dietary management of irritable bowel syndrome, ulcerative colitis, and pouchitis. VSL#3-DS, which contains a greater concentration of bacteria than VSL#3, is also designated as a medical food for the management of the same conditions. VSL#3-DS requires a physician's prescription because of its higher potency ("VSL#3" 2012).

Cosmetic: Cosmetics are not subject to premarket approval by the FDA, nor are cosmetics required to register with the FDA or even report cosmetic-related injuries to the FDA. The FDA does not have the power to force a cosmetic company to recall a tainted or impure product from the market. The FDA regulations "prohibit the marketing of adulterated or misbranded cosmetics...whether they result from ingredients, contaminants, processing, packaging, or shipping and handling" (FDA 2005). The FDA also has authority over the labeling of cosmetics, in that they cannot be "improperly labeled" or "deceptively packaged" (Ibid.). In addition, the FDA requires that the product include in its packaging an ingredient list.

The Federal Trade Commission: The Federal Trade Commission (FTC) and the FTC Act, sections 5 and 12, also regulate the kinds of claims that may be made about products, including claims related to over-the-counter drugs, food, devices, and cosmetics. Section 5 of the FTC Act prohibits "unfair or deceptive acts or practices in or affecting commerce." Section 12 prohibits "disseminating or causing the dissemination of a false advertisement in commerce for the purpose of inducing, or that is likely to induce, the purchase of any food, drug, device, service or cosmetics." For example, the FTC required Nestlé, the manufacturer of "Kid Essentials," to change their advertising and labeling. They marketed a probiotic straw and drink for children. The packaging and advertisements claimed that the product could prevent upper respiratory infections, protect against the cold and flu, reduce school absences, and reduce the duration of acute diarrhea in children (Federal Trade Commission 2010).

In sum, current regulatory structure has no clear and useful mechanism to require systematic scientific study of probiotics or phages and their effect on the human microbiome. On the one hand, the current regulatory structure allows manufacturers to market their products without providing either physicians or consumers with the research data to support claims about their efficacy. Thus, manufacturers who have established a market for their products have no incentive to conduct studies that might demonstrate that their products are less effective than consumers believe them to be. On the other hand, the regulations have no obvious mechanism for encouraging rigorous and accurate investigation of probiotics and phages. By claiming that they are safe, manufacturers are reluctant to treat probiotics as drugs. At the same time, categorizing probiotics and phages as something other than drugs leaves them with no regulatory mechanism to provide assurance of their safety and efficacy. As our understanding of the role of the microbiome increases, the need for addressing this lacuna will become more pressing.

Conclusion

Many medical conditions are not adequately addressed with currently available treatments. Without human subject research, improvements in medical treatment will not be achieved. We and our loved ones shall remain vulnerable to avoidable death, disease, injury, pain, suffering, and disability.

Biological differences, for the most part, have neither biological nor moral significance in the context of biomedical research. Biophysiological processes in one human are very much like those in others. This means that knowledge gained from studying one individual is likely applicable to another, and vice versa. And ethically, there are no physical differences that are so significant as to mark some of us as research material and exempt others from the moral responsibility of contributing to the communal promotion of biomedical science.

Improving the benefits that medicine can provide can only be achieved by studying our bodies. Whereas other social obligations can be fulfilled in other ways (e.g., by writing a check, by sending a stand-in), at a certain point in biomedical research, there is nothing that we can substitute for us. Study involves some sacrifice of our flesh, our privacy, our safety, our comfort, and our time. Because these basic goods are precious to everyone, basic moral principles of equality, universalization, and mutual love require us to give of ourselves as we would wish to receive from others. The fragility of our bodies, the invasiveness of research, our emotional and genetic interrelatedness, the lack of an adequate alternative, and the commonality of the desire to benefit from medical knowledge inspire us to adopt a new perspective on research participation. Solidarity with our fellow humans tells us that we should each be willing to do our fair share and participate in advancing biomedical science (Rhodes 2005).

Almost everyone and almost all of our loved ones have medical needs at some point in their lives. Because researchers sometimes need to study subjects who are ill, subjects from different genetic groups, and entire populations, the participation of all kinds of people is needed in the enterprise. Some biomedical advances related to the human microbiome cannot be achieved without widespread participation. Some advances would proceed slowly without general participation. Research on the human microbiome, the role of the microbiome in disease resistance and susceptibility, and the development of probiotic and bacteriophage tools for combating illness presents important new opportunities for advancing biomedical science. The opportunity to advance our understanding of the human body and its microbial environment would be squandered without widespread participation in the studies that are on the horizon.

Microbiome research offers huge possibilities for progress in our understanding of the interactions between us and the bacteria, viruses, and fungi that live in us and on us. Looking at this vast new domain in the life sciences should alert us to the need for reflectively examining our ideas about the ethical conduct of human subject research and encourage us to change indefensible attitudes toward research participation. As Claude Bernard explained, "When we meet a fact which contradicts a prevailing theory, we must accept the fact and abandon the theory, even when the theory is supported by great names and generally accepted" [Bernard [1865] 1949, 164]. Now is the time to give up the view that there is no duty to participate in research. We need to see ourselves and our fellow humans as cooperating autonomous individuals who each has the responsibility to respect one another and collaborate in important projects that advance our shared interests. It is time to acknowledge the social purpose of research and design policies in that light.

Regarding research from this perspective will encourage clinicians to view treatment and research as going hand in hand. It should also allow them to see the artificiality and the conceptual flaws in the regulation-constructed definitions of "research" and treatment "innovation." This manufactured distinction allows most anything to be done to "patients" as "treatment" while significantly restricting what may be done to "subjects" who are participants in "research." A treatment approach that also embraces research would allow clinician-researchers to store and identify specimens and records for later use in projects that have not yet been conceived, to follow up with identifiable research participants to help confirm (or disconfirm) hypotheses, and to recontact participants when they are needed to help further future research goals. This approach would allow researchers and clinicians to urge patients and others to participate in research. This shift in attitude would also allow us to appreciate that we have a moral duty to do our fair share. Those who refuse to participate in research have to justify their refusal, at least to themselves. This change puts the onus on the opposite side of where it has been, and that is quite a radical difference.

Biomedical research on the human microbiome introduces a radical shift in how we conceptualize research ethics. Instead of seeing research as an activity outside of our moral world, looking through the lens of the HMP allows us to see that scientific study is an integral piece of the social fabric. This new perspective encourages a critical examination of the ethical rules and constraints on human subject research and allows us to liberate research ethics from its well-intentioned ideology. Seeing research participants as people who are capable of making choices that reflect their own values allows us to stop viewing them as vulnerable subjects who require paternalistic protection. It also allows policymakers to move away from a narrow focus on informed consent as the primary ethical value to be considered in research oversight. A shift in that direction would encourage policymakers to fully appreciate that a number of important issues have to be considered. The ethical assessment of a study should turn on judgments of the risks involved and when and for whom these risks would be reasonable. The ethical conduct of a study may require IRB review that scrutinizes the research plan, the informed consent process, the actual conduct of the study, and the communication between researchers and research participants.

A new perspective on research ethics would replace a one-size-fits-all approach to research ethics with one that is sensitive to the context of the project, the risks, and the burdens involved. In sum, viewing research ethics in the context of the microbiome studies on the horizon tells us that reasonable research policy begins with an evaluation of the scientific merits and an assessment of risks and burdens for participants. This new perspective would refocus the role of IRBs toward protecting the trust and trustworthiness of biomedical research.

Policy Recommendation

- Our statements and recommendations about research are largely compatible with current regulations. In some institutions, however, IRBs may interpret the regulations narrowly and impose unreasonable demands that inhibit research. Burdensome requirements should be scrutinized and either justified, modified, or eliminated.
- Research ethics regulations and policies have primarily focused on avoiding risks and providing benefits to research participants. They should also give significant weight to advancing knowledge and providing benefits to future patients.
- The level of research oversight should vary in relation to the level of anticipated risk. Studies with very minimal risks do not require the same degree of oversight as do studies that involve significant risks. Regulations should be adjusted to reflect these differences.
- Regulations should distinguish a subcategory of *de minimis* risk research from other minimal-risk research. The term *de minimis risk* indicates a negligible risk that involves only hard-to-imagine and extremely unlikely minimal risks.

- In studies that involve only a *de minimis* risk, informed consent should not be required.
- Once samples are collected, microbiome biobank and sample bank studies involve only *de minimis* risk. Informed consent should not be required for studies using previously banked samples.
- Microbiome studies that do not involve directly interfering with participants' bodies (e.g., sampling effluent from a community) involve only *de minimis* risk. When samples are taken for *de minimis* risk surveillance, quality assurance, or quality improvement, informed consent should not be required.
- In the course of clinical care, biological samples are routinely collected for analysis and remaining material is often discarded. The risks involved of using otherwise discarded samples are only *de minimis*. Institutions should adopt an opt-out policy for their use in research. When obtaining agreement for sample use would be unfeasible or unreasonably burdensome, informed consent should not be required.
- In some states and in some institutions, surrogate consent for research is not accepted or restricted. In some studies, the value of collecting samples from people who cannot consent could be significant. Surrogate consent for microbiome research and biobank and sample bank participation should be accepted as sufficient for specimen collections that involve only minimal or *de minimis* risk.
- Some therapeutic microbiome research, for example, clinical trials of probiotics or phages, may involve greater-than-minimal risk. When surrogate consent would be sufficient for an intervention as an innovative treatment, it should also be adequate for clinical research on the efficacy and risks of the intervention.
- In medical treatment, medical education, and biomedical research, information about people should be treated according to existing standards of "confidentiality." Information should be shared only on a need-to-know basis and in accordance with professional safeguards of confidentiality. No additional "privacy" restrictions should be required.
- Microbial information generated by health care and human microbiome research should be legally protected against discrimination in health insurance and employment and safeguarded from criminal and immigration investigations. Biobanks and sample banks should be protected against subpoena.
- When research review boards determine that the risks involved in a study are reasonable and have an acceptable risk–benefit ratio, incentives sufficient to ensure timely and efficient completion of a study should be allowed.

- The current regulatory structure has no clear and useful mechanism for requiring systematic scientific study of probiotics and phages and their effect on the human microbiome. A new mechanism to regulate probiotics and phages is required.

References

Abbott, A. 1999. 'Strengthened' Icelandic Bioethics Committee comes under fire. *Nature* 400: 62.

Ballantyne, A. 2008. Benefits to Research Subjects in International Trials: Do they reduce exploitation or increase undue inducement? *Developing World Bioethics* 8: 178–191.

Beecher, H. K. 1966. Ethics and clinical research. *New England Journal of Medicine* 274(24): 1354–1360.

Bernard, C. 1927. *An introduction to the study of experimental medicine* (H.C. Greene, Trans.). New York: Macmillan & Co., Ltd (Original work published in 1865).

Birren, B. W. 2011. U.S. Department of Health and Human Services, National Institutes of Health. Defining the human microbiome (NIH Publication No. 1U54HG004969-01). Washington, DC: Retrieved from http://projectreporter.nih.gov/project_description.cfm?projectnumber=1U54HG004969-01. Accessed January 24, 2011.

Blaser, M. J. 2012. Heterogeneity of Helicobacter pylori. *European Journal of Gastroenterology and Hepatology* 9(Suppl 1):S3–S6; discussion S6–S7.

Borrell, B. 2012. Phage factor. Interview of Fischetti, V. *Scientific American* 307(2): 80–83.

Breen, G., and M. Loughran. 2005. The genographic project: Mapping the human story. [Fact Sheet].

Code of Federal Regulations. 1991. Federal Policy for the Protection of Human Subjects The Federal Policy for the Protection of Human Subjects, (the "Common Rule") Title 45, Part 46. Protection of Human Subjects. Rockville, MD: U.S. Department of Health and Human Services. http://www.wma.net/en/30publications/10policies/b3/17c.pdfhttp://www.hhs.gov/ohrp/humansubjects/guidance/45cfr46.html Accessed April 3, 2013.

Conference Participants, 2001. Conference on Ethical Aspects of Research in Developing Countries Ethics: fair benefits for research in developing countries. Science 2002; 298:2133–4.

Cordina, C., I. Shaikh, S. Shrestha, and J. Camilleri-Brennan. 2011. Probiotics in the management of gastrointestinal disease: analysis of the attitudes and prescribing practices of gastroenterologists and surgeons. *Journal of Digestive Diseases* 12: 489–496.

Councils for International Organizations of Medical Sciences. 2002. *International Ethical Guidelines for Biomedical Research Involving Human Subjects (CIOMS Guidelines).* http://www.codex.uu.se/texts/international.html. Accessed April 3, 2013.

De Zoysa, I. J. Elias, and M. E. Bently. 1998. Ethical challenges in efficacy trials of vaginal microbiocides for HIV prevention. *American Journal of Public Health* 88: 571–575.

Drivsholm, T., L. F. Eplov, M. Davidsen, T. Jorgensen, H. Ibsen, H. Hollnagel, and K. Borch-Jonsen. 2006. Representativeness in population-based studies: a detailed description of non-response in a Danish cohort study. *Scandinavian Journal of Public Health* 34: 623–631.

Emanuel, E., D. Wendler, and C. Grady. 2000. What makes clinical research ethical? *Journal of the American Medical Association* 283: 2701–2711.

Emanuel, E. J. 2004. Ending concerns about undue inducement. *Journal of Law, Medicine and Ethics* 32: 100–105.

Emanuel, E. J., E. C. Xolani, and A. Herman. 2005. Undue inducement in clinical research in developing countries: is it a worry? *Lancet* 366: 336–340.

Emanuel, E. J., E. Abdoler, and L. Stunkel. 2009. Research ethics: How to treat people who participate in research. Retrieved from http://www.bioethics.nih.gov/education/FNIH_BioethicsBrochure_WEB.PDF

Erisman, J. W., M. A. Sutton, J. Galloway, Z. Klimont, and W. Winiwarter. 2008. How a century of ammonia synthesis changed the world. *Nature Geoscience* 1: 636–639.

Estonian Genome Center of the University of Tartu, Estonian Genome Center. 2009. Retrieved from http://www.geenivaramu.ee/index.php?id=462

Faden, R. R., and T. L. Beauchamp. 1986. *A history and theory of informed consent.* New York: Oxford University Press.

Federal Trade Commission. 2010. In the matter of Nestle healthcare nutrition (FTC File No. 092 3087). http://www.ftc.gov/os/caselist/0923087/index.shtm. Accessed July 22, 2011.

Fettweis, J. M., J. P. Alves, J. F. Borzelleca, J. P. Brooks, C. J. Friedline, Y. Gao, X. Gao, P. Girerd, M. D. Harwich, S. L. Hendricks, K. K. Jefferson, V. Lee, H. Mo, M. C. Neale, F. A. Puma, M. A. Reimers, M. C. Rivera, S. B. Roberts, M. G. Serrano, N. U. Sheth, J. L. Silbert, L. Voegtly, E. C. Prom-Wormley, B. Xie, T. P. York, C. N. Cornelissen, J. L. Strauss, L. J. Eaves, and G. A. Buck. 2011. The vaginal microbiome: disease, genetics and the environment, *Nature Precedings*, Posted March 9.

Fierer, N., M. Hamady, C. L. Lauber, and R. Knight. 2008. The influence of sex, handedness, and washing on the diversity of hand surface bacteria. *Proceedings of the National Academy of Sciences of the United States of America* 105(46): 17994–17999.

Fierer, N., C. L. Lauber, N. Zhou, D. McDonald, E. K. Costello, and R. Knight. 2010. Forensic identification using skin bacterial communities. *Proceedings of the National Academy of Sciences of the United States of America* 107(14), 6477–6481.

Fischetti, V. A. 2006. Using phage lytic enzymes to control pathogenic bacteria. *BMC Oral Health* 6(Suppl 1): S16.

Fischetti, V. A., D. Nelson, and R. Schuch. 2006. Reinventing phage therapy: are the parts greater than the sum? *Nature Biotechnology* 24: 1508–1511.

Food and Agriculture Organization (FAO) of the United Nations and World Health Organization (WHO) Working Group. 2002. Joint FAO/WHO Working Group Report on Drafting Guidelines for the Evaluation of Probiotics in Food. London, Ontario, Canada, April 30 and May 1. http://www.who.int/foodsafety/fs_management/en/probiotic_guidelines.pdf. Accessed April 12, 2013.

Ford, J. G., M. W. Howerton, G. Y. Lai, T. L. Gary, S. Bolen, M. C. Gibbons, J. Tilburt, C. Baffi, T. P. Tanpitukpongse, R. F. Wilson, N. R. Powe, and E. B. Bass. 2008. Barriers to recruiting underrepresented populations to cancer clinical trials: A systematic review. *Cancer* 112: 228–242.

Foster, M. W. 2004. Integrating ethics and science in the International HapMap Project. *Nature Reviews Genetics* 5: 467–475.

Freeman, T. B., D. E. Vawter, P. E. Leaverton et al. 1999.Use of placebo surgery in controlled trials of a cellular based therapy for Parkinson's disease. *New England Journal of Medicine* 341: 988–91.

Gibson, G. R., and M. B. Roberfroid. 1995. Dietary modulation of the human colonic microbiota. Introducing the concept of prebiotics. *Journal of Nutrition* 125: 1401–1412.

Gong M., L. Richardson, R. Rhodes, G. Winkel, and J. Silverstein. 2010. Surrogate consent for research involving adults with impaired decision-making: survey of institutional review boards' practices. *Critical Care Medicine* 38(11): 2146–2154.

Gregory, J. (1770) 1998. *Observations on the Duties and Offices of a Physician, and on the Method of Presecuting Enquiries in Philosophy* (London: W. Strahan and T. Cadell). Reprinted in *John Gregory`s Writings on Medical Ethics and Philosophy* (Ed. Laurence B. McCullough). Dordrecht: Kluwer Academic Publishers.

Greely, H. T. 2001. Informed consent and other ethical issues in human population genetics. *Annual Review of Genetics* 35: 785–800.

Gulcher, J. R., and K. Stefansson. 1999. Ethics of population genomics research. *Nature* 400: 307–308.

Grice, E., et al. 2009. Topographical and temporal diversity of the human skin microbiome. *Science* 324(5931): 1190–1192.

Gulcher, J. R., and K. Stefansson. 1999. The Icelandic healthcare database and informed consent. *New England Journal of Medicine* 342: 1827–1830.

Haakonssen, L. 1997. *Medicine and Morals in the Enlightenment: John Gregory, Thomas Percival and Benjamin Rush.* Atlanta, GA: Rodopi B.V., Amsterdam.

Halpern, S. D., J. H. T. Karlawish, D. Casarett, J. A. Berlin, and D. A. Asch. 2004. Empirical assessment of whether moderate payments are undue or unjust inducements for participation in clinical trials. *Archives of Internal Medicine* 164: 801–803.

Harmon, A. 2006. DNA gatherers hit snag: tribes don't trust them. *New York Times,* December 10: A1.

Hawkins, J., and E. J. Emanuel. 2005. Clarifying confusions about coercion. *Hastings Center Report* 35: 16–19.

Hornsby, John A., and R. E. Schmidt. 1913. *The modern hospital.* Philadelphia, PA: W.B. Saunders.

Human Epigenome Consortium. Human epigenome project. Retrieved from http://www.epigenome.org/index.php?page=project.

Human Variome Project. 2011. Retrieved from http://www.humanvariomeproject.org/index.php/about

International HapMap Project, 2002. Retrieved from http://hapmap.ncbi.nlm.nih.gov/index.html.en

Jansen, L. A. 2005. A closer look at the bad deal trial: Beyond clinical equipoise. *Hastings Center Report* 35: 29–36.

Jonas, H. 1969. Philosophical reflections on experimenting with human subjects. *Daedalus* 98(2): 219–247.

Jones J. H. 1993. *Bad blood : The Tuskegee Syphilis Experiment.* New York: The Free Press.

Kinross, J. M., A. C. von Roon, E. Holmes, A. Darzi, and J. K. Nicholson. 2008. The human gut microbiome: implications for future health care. *Current Gastroenterology Reports* 10: 396–403.

Koenig, J. E., A. Spor, N. Scalfone, A. D. Fricker, J. Stombaugh, R. B. Knight, L. T. Angenent, and R. E. Ley. 2010. Succession of microbial consortia in the developing infant gut microbiome. *Proceedings of the National Academy of Sciences of the United States of America* 108: 4578–4585.

Kopelman, L. M. 2012. Health care reform and children's right to health care: a modest proposal. In R. Rhodes, M. P. Battin, and A. Silvers (eds.), *Medicine and Social Justice: Essays on the Distribution of Health Care* (pp. 335–346). New York: Oxford University Press.

Lederberg, J., and A. T. McCray. 2001. 'Ome sweet 'omics—a geneological treasure of words. *Scientist* 15(7): 8.

Lederer, S. 1995. *Subjected to Science: Human Experimentation in America Before the Second World War*. Baltimore: Johns Hopkins University Press.

Lederer, S. E., and M. A. Grodin. 1994. Historical overview: pediatric experimentation. In M. A. Grodin and L. H. Glantz (eds.), *Children as Research Subjects: Science, Ethics and Law* (pp. 3–22). New York: Oxford University Press.

Lee, Y. K., and S. Salminen. 2009. *Handbook of Probiotics and Prebiotics*. Hoboken, NJ: John Wiley and Sons.

Macklin, R. 1981. 'Due' and 'undue' inducements: on passing money to research subjects. *IRB: Ethics and Human Research* 3(5): 1–6.

Macklin, R. 1999. The ethical problem with sham surgery in clinical research. *New England Journal of Medicine* 341: 992–996.

Macklin, R. 2004. *Double Standards for Medical Research in Developing Countries*. Cambridge: Cambridge University Press.

McGuire, A. L., J. Colgrove, S. N. Whitney, C. M. Diaz, D. Bustillos, and J. Versalovics. 2008. Ethical, legal, and social considerations in conducting the Human Microbiome Project. *Genome Research* 18: 1861–1864.

McGuire, A. L., and J. R. Lupski. 2010. Personal genome research: what should the participant be told? *Trends in Genetics* 26(5): 199–201.

Miller, F. G., and Brody, H. 2002. What makes placebo-controlled trials unethical? *The American Journal of Bioethics* 2: 3–9.

Miller, F. G., and Brody, H. 2003 A critique of clinical equipoise: Therapeutic misconception in the ethics of clinical trials. *Hastings Center Report* 33: 19–28.

Moros, D. A., and Rhodes, R. 2010. Privacy overkill, *The American Journal of Bioethics* 10: 12–15.

Mfutso-Bengo, J., Mayisie, F., Molyneux, M., Ndeble, P., and Chilungo, A. 2008. Why do people refuse to take part in biomedical research studies? Evidence from a resource-poor area. *Malawi Medical Journal* 20: 57-63.

National Commission for the Protection of Human Subjects of Biomedical and Behavioral Research. 1979. *The Belmont Report: Ethical Principles and Guidelines for the Protection of Human Subjects of Research*. Washington, DC: U.S. Government Printing Office. http://hhs.gov/ohrp/humansubjects/guidance/belmont.html. Accessed July 21, 2011.

Nuremberg Code. II Trials of War Criminals Before the Nuremberg Military Tribunals Under Control Law No. 10, 181-83. U.S. Government Printing Office. 1946–1949. http://www.hhs.gov/ohrp/archive/nurcode.html Accessed, April 3, 2013.

NIH Clinical Trials: Various Factors Affect Patient Participation. Washington, DC: US General Accounting Office; September 1999. Document GAO/HEHS-99-182.

Norwegian Institute of Public Health, Norwegian Mother and Child Cohort Study. 2008. Retrieved from http://www.fhi.no/eway/default.aspx?pid=238&trg=MainArea_5811&MainArea_5811=5903:0:15,3046:1:0:0:::0:0

Offices of the Secretary, DHHS, FDA. 1996. Protection of Human Subjects: Informed Consent and Waiver of Informed Consent Requirements in Certain Emergency Circumstances: Final Rule. Fed Reg. 61, 51490-533.

Osborne, L. 2000. A Stalinist antibiotic alternative. *New York Times,* February 6.

Park, W. 1931. The history of diphtheria in New York City. *American Journal of Diseases of Children* 42(6): 1439–1445.

Participants in the 2001 Conference on Ethical Aspects of Research in Developing Countries. 2004. Moral standards for research in developing countries: From "Reasonable Availability" to "Fair Benefits." *Hastings Center Report* 34: 17–26.

Percival, T. (1803). 1985. Medical ethics: A code of institutes and precepts, Adapted to the Professional Conduct of Physicians and Surgeons by Thomas Percival, M. D. (Ed. Edmund Pellegrino). Birmingham, AL: Classics of Medicine Library.

Plus News. Oct. 2007. Microbiocide trials: What's in it for the participants? In Trials and Tribulations of HIV Prevention Research. Retrieved from http://www.plusnews.org/IndepthMain.aspx?InDepthID=64&ReportID=74136

Reverby, S. M. 2009. *Examining Tuskegee: The Infamous Syphilis Study and Its Legacy.* Chapel Hill, NC: University of North Carolina Press.

Rose, H. 2003 An ethical dilemma: the rise and fall of human genomics–the model biotech company? *Nature* 425(6954): 123–124.

Rescha, G., P. Moreillon, and V. A. Fischettia. 2011. A stable phage lysin (Cpl-1) dimer with increased antipneumococcal activity and decreased plasma clearance. *International Journal of Antimicrobial Agents* 38(6): 516–521.

Rhodes R. 2005. Rethinking Research Ethics. *American Journal of Bioethics* 5(1): 7–28.

Scanlon, T. M. 1998. *What We Owe to Each Other.* Cambridge, MA: Harvard University Press.

Schmitz, J. E., M. C. Ossiprandi, K. R. Rumah, and V. A. Fischetti. 2011. Lytic enzyme discovery through multigenomic sequence analysis in clostridium perfringens. *Applied Microbiology and Biotechnology* 89(6): 1783–1795.

Stern, A., E. Mick, I. Tirosh, O. Sagy, and R. Sorek. 2012. CRISPR targeting reveals a reservoir of common phages associated with the human gut microbiome. *Genome Research* 22(10): 1985–1994.

Temple, R., and S. S. Ellenberg. 2000. Placebo-controlled trials and active-control trials in the evaluation of new treatments. Part 1: Ethical and scientific issues. *Annals of Internal Medicine* 133: 455–463.

Trust Sanger Institute, Cancer Genome Project. 2009. http://www.sanger.ac.uk/genetics/CGP/

Turnbaugh, P. J., Ley, R. E., Hamady, M., Fraser-Liggett, C. M., Knight, R., and Gordon, I. J. 2007. The Human Microbiome Project. *Nature* 449: 804–810.

Turnbaugh, P. J., M. Hamady, T. Yatsunenko, B. L. Cantarel, A. Duncan, R. E. Ley, et al. 2008. A core gut microbiome in obese and lean twins. *Nature* 457: 480–484.

U.S. Food and Drug Administration. 2002. Guidance for industry: structure/function claims, small entity compliance guide. January 9. http://www.fda.gov/Food/GuidanceRegulation/GuidanceDocumentsRegulatoryInformation/DietarySupplements/ucm103340.htm. Accessed July 22, 2011.

U.S. Food and Drug Administration. 2003. Claims that can be made for conventional foods and dietary supplements. http://www.fda.gov/Food/IngredientsPackagingLabeling/LabelingNutrition/ucm111447.htm

U.S. Food and Drug Administration. 2004. GMPs—section one: current food good manufacturing practices. http://www.fda.gov/Food/GuidanceRegulation/CGMP/ucm110907.htm. Accessed April 12, 2013.

U.S. Food and Drug Administration. 2005. FDA authority over cosmetics. http://www.fda.gov/Cosmetics/GuidanceComplianceRegulatoryInformation/ucm074162.htm. Accessed July 22, 2011.

U.S. Food and Drug Administration. 2007. Guidance for industry: frequently asked questions about medical foods. http://www.fda.gov/Food/GuidanceRegulation/GuidanceDocumentsRegulatoryInformation/MedicalFoods/ucm054048.htm. Accessed April 12, 2013.

U.S. Food and Drug Administration. 2009. Q&A on Dietary Supplements. http://www.fda.gov/Food/DietarySupplements/QADietarySupplements/default.htm#what_is. Accessed April 12, 2013.

U.S. Food and Drug Administration. 2010.Office of Drug Evaluation IV: What We Do. http://www.fda.gov/AboutFDA/CentersOffices/OfficeofMedicalProductsandTobacco/CDER/ucm106342.htm. Accessed May 10, 2013.

U.S. Department of Health and Human Services. 2011. Human Subjects Research Protections: Enhancing Protections for Research Subjects and Reducing Burden, Delay, and Ambiguity for Investigators. http://www.hhs.gov/news/press/2011pres/07/20110722a.html. Accessed July 22, 2011.

U.S. Food and Drug Administration. 2011a. Generally Recognized as Safe (GRAS). http://www.fda.gov/Food/IngredientsPackagingLabeling/GRAS/default.htm. Accessed April 12, 2013.

U.S. Food and Drug Administration. 2011b. Regulation of nonprescription products. http://www.fda.gov/AboutFDA/CentersOffices/OfficeofMedicalProductsandTobacco/CDER/ucm093452.htm. Accessed April 12, 2013.

U.S. Food and Drug Administration. n.d. Federal Food, Drug and Cosmetic Act SEC. 201. [21 U.S.C. 321] FEDERAL FOOD, DRUG, AND COSMETIC ACT, CHAPTER II—DEFINITIONS 1. http://www.gpo.gov/fdsys/pkg/USCODE-2010-title21/html/USCODE-2010-title21-chap9-subchapII.htm. Accessed April 12, 2013.

U.S. Food and Drug Administration. n.d. Federal Food, Drug and Cosmetic Act SEC. 201. [21 U.S.C. 321] CHAPTER II—DEFINITIONS 1. http://www.gpo.gov/fdsys/pkg/USCODE-2010-title21/html/USCODE-2010-title21-chap9-subchapII.htm. Accessed April 12, 2013.

U.S. Food and Drug Administration/CFSAN. 2007. Agency Response Letter, GRAS Notice No. 000198.

U.S. Department of Energy Office of Science, Human Genome Program. Ethical, legal and social issues (ELSI) (DOE Microbial Genomics). Washington, DC: Retrieved from http://www.ornl.gov/sci/techresources/Human_Genome/elsi/elsi.shtml

Veatch, R. M. (1987). The patient as partner: A theory of human-experimentation ethics. Bloomington: Indiana University Press.

Ventura, M., T. Sozzi, F. Turroni, D. Matteuzzi, and D. van Sinderen. 2011. The impact of bacteriophages on probiotic bacteria and gut microbiota diversity. *Genes & Nutrition* 6(3): 205–207.

"VSL3." 2012. Sigma-Tau Pharmaceuticals, Inc. http://www.vsl3.com/. Accessed August 6, 2012.

Wade, N. 2010. Out of bankruptcy, genetics company drops drug efforts. *New York Times*, January 21, p. B2. http://www.nytimes.com/2010/01/22/business/global/22gene.html. Accessed April 12, 2013.

Wilson, I. D. 2009. Drugs, bugs, and personalized medicine: pharmacometabonomics enters the ring. *Proceedings of the National Academy of Sciences of the United States of America* 106(34): 14187–14188.

World Health Organization. 2003. Manual for the laboratory identification and antimicrobial susceptibility testing of bacterial pathogens of public health importance in the developing world. WHO/CDS/CSR/RMD/2003.6. http://www.who.int/csr/resources/publications/drugresist/WHO_CDS_CSR_RMD_2003_6/en/. Accessed April 12, 2013.

World Medical Association Declaration of Helsinki. 1964. http://www.wma.net/en/30publications/10policies/b3/17c.pdf Accessed, April 3, 2013.

Zalloua, P. A., D. E. Platt, M. El Sibai, J. Khalife, N. Makhoul, M. Haber, Y. Xue, H. Izaabel, E. Bosch, S. M. Adams, E. Arroyo, A. M. López-Parra, M. Aler, A. Picornell, M. Ramon, M. A. Jobling, D. Comas, J. Bertranpetit, R. S. Wells, and C. Tyler-Smith, Genographic Consortium. 2008. Identifying genetic traces of historical expansions: Phoenician footprints in the Mediterranean. *American Journal of Human Genetics* 83(5): 633–642.

Zaura, E., B. J. F. Keijser, S. M. Huse, and W. Crielaard. 2009. Defining the healthy "core microbiome" of oral microbial communities. *BMC Microbiology* 9: 259.

6

Biobanks and the Human Microbiome

Abraham P. Schwab, Barbara Brenner, Joseph Goldfarb, Rochelle Hirschhorn, Sean Philpott

Introduction

Research into the human microbiome relies heavily on the use of biobanks, sample banks, and databanks.[1] The existing regulations that govern human subject research were not designed with biobanks in mind, and, at present, they cannot be meaningfully applied to human microbiome biobanks.[2] Thus, the ethical, social, and legal implications that arise as biobanks and databanks are established and maintained demand attention.

We begin this chapter by describing the functions and activities of biobanks. These vary across types of biobanks and with each biobank's purpose. From there we investigate the obstacles to applying existing federal regulations to the use of microbiome biobanks for medical research. We also identify some of the challenges that face microbiome biobanks because they require broad participation and because of the usefulness of comparing biobanked samples from across the globe. Finally, we discuss the role that community consultation can and should play in the ethical governance of microbiome biobanks.

What Is a Biobank?

Biobanks, also known as biorepositories, are collections of biological specimens and associated health data used for research purposes. The practice of

[1] It is worth noting that strategies for isolating and culturing the microbiome are still being developed. Less than two years ago, isolating and culturing the bacteria comprising the human microbiome was not yet possible. By the end of 2010, strategies had been developed that enable scientists to isolate and culture many of these bacteria. Often these strategies are labor intensive. (These developments are also discussed in Chapter 1: The Human Microbiome: Science, History and Research.)

[2] It should be noted that the U.S. Department of Health and Human Services (HHS) has recently completed an open forum for discussing possible changes to the Common Rule.

storing biological samples has been around for centuries. The widespread use of stored samples in research has blossomed in the past two decades, and biobanked samples are playing an increasingly important and valuable role in biomedical research (National Bioethics Advisory Commission 1999). Biobanks are particularly useful for research into the susceptibility to various human diseases, both transmitted illnesses and genetic disorders, as well as for the identification of markers to predict disease susceptibility, progression, and response to treatment.

Specimens stored in biobanks can include tissue samples, body fluids (e.g., blood, serum, plasma, and saliva), waste products (e.g., hair, nail clippings, urine, and feces), cells (e.g., from cheek and skin swabs), and genetic material (both DNA and RNA). The types of specimens collected, and the amount of associated clinical information stored, varies with the biobank's purpose and scope. For instance, one of the largest proposed biobanks (named simply the UK Biobank) aims to collect tissue and fluid samples from almost 500,000 British residents. Designed to serve as "a major resource that can support a diverse range of research intended to improve disease prevention, diagnosis, and treatment of illness and the promotion of health throughout society," the UK Biobank will collect tissue and fluid specimens and link these specimens to the donor's full medical history, as well as additional lifestyle and environmental data (UK Biobank Ethics and Governance Framework 2007). By contrast, the Oregon Health Science University's Layton Aging & Alzheimer's Disease Center maintains a far smaller and more focused biobank. It collects and stores tissue and associated autopsy records from 58 families with a history of Alzheimer's disease (Payami et al. 1996). Thus, biobanks can range from collections of specimens from a large number of individuals in different states of health to focused specimen banks that collect specific samples and data from individuals with a particular disease or disorder.

The collection of biological specimens stored in biobanks often occurs as part of a medical intervention during which tissues or fluids are already being sampled for diagnostic or treatment purposes (Riegman and van Veen 2011). Following surgical procedures, for example, portions of surgically excised tissue are routinely stored for future clinical or research purposes. Often, patients[3] agree to biobank these specimens as part of the surgical consent process. For clinical and research purposes, however, the present and future value of the sample requires linking the specimen to pertinent medical data.

---

[3] There is an ongoing disagreement about what to call those individuals who agree to take part in human subject research. In many cases, they begin as patients, but once they enter the research protocol, clearly, they are no longer simply that. *Research subjects* has long been in use, but it is associated with ethically tarnished medical research (e.g., Tuskegee, Willowbrook, etc.). Recently, the more preferred term is *research participant*. Throughout, for individuals who give specimens and information to biobanks, we will use *donor*. This serves to distinguish individuals who provide specimens and information to biobanks from individuals involved in clinical trials.

Thus, specimens are most often stored in a way that links them, directly or indirectly, to identifiable personal health information like diagnoses, prognoses, and treatments. Advances in information technology enable biobanks to link samples and other patient data and enable researchers to search, or mine, this information for a variety of purposes. Epidemiologists, for example, might use a linked database to determine the frequency of a disease in a particular population or to predict its spread in the general population (Brandt and Probst-Hensch 2007; Ioannidis and Adami 2008). Clinicians might use linked data and stored specimens to develop a new model of a disease etiology and progression (Goldacre et al. 2000; Ioannidis and Adami 2008; Riegman, Dinjens, and Oosterhuis 2007). Biobanks thus provide scientists and clinicians with a new and powerful set of research tools that directly link tissue, fluid, microbiome, and genetic samples with vast compilations of medical data.

In addition to samples that are obtained through routine medical procedures, biological specimens may also be obtained from individuals who consent either to participate in a research study or to donate to a biobank. In the case of microbiome research, samples will be obtained primarily from skin and various body cavities. Many of these samples will be gathered using procedures (e.g., skin swabs, oral swabs) that involve negligible (or *de minimis*) physical risk (Rhodes et al. 2011). Collecting some microbiome samples, however, may involve some small degree of risk, such as when an extra bit of material is taken from the gut during a diagnostic procedure.

Once a sample has been taken, biobanks preserve both the sample itself and an electronic coding of the sample's genome. Biobanks also utilize various data management techniques for associating information about the donor and the donor's medical history with the sample or disassociating the sample from information about the donor (Brandt and Probst-Hensch 2007; Ioannidis and Adami 2008; Riegman et al. 2007). Samples may be stored as anonymous, unlinked anonymized, linked anonymized, coded (i.e., de-identified), or identified samples. Because it is possible to reconnect a sample to an individual through DNA analysis when you have or can obtain another identified sample for comparison, only archeological samples are truly anonymous samples (Brothers and Clayton 2010). Anonymized samples are those that are stripped of the information that would allow possible identification of the donor. When samples are delinked, the stripping is irreversible—there is no way to relink the sample and/or information to the donor. By contrast, if the anonymized sample is linked, the donor information can be reconnected to the sample by the biobank administrator (but not the researcher). Coded (or de-identified) samples are similar to linked anonymized samples, except that the linking mechanism may be provided to researchers along with the sample or other information. Identifiable samples have not been stripped of identifiable data at all (Elger and Caplan 2006).

Human Microbiome Biobanks

The ethical issues and logistical challenges arising with the use of microbiome biobanks vary with the nature of the research. Some microbiome researchers, for example, are interested in determining whether all human beings share a single identifiable "core" microbiome (Shade and Handelsman 2012). Other researchers are interested in how the microbiome is affected by individual genotypes and environmental exposures, and whether changes in the microbiome over time cause, or are predictive of, human disease (Turnbaugh et al. 2007). Some biobanks may collect and store different types of samples, including cultures of commensal or symbiotic microorganisms from different sites on or within the human body. Some of these microorganisms have yet to be identified or may be difficult to culture in the laboratory, thus presenting a novel problem for biobank managers.

To examine how the human microbiome evolves over time or in response to age, environment, diet, lifestyle, and disease, it may also be necessary to collect multiple samples from individual donors over time. Depending on the nature and number of samples collected, such a prospective cohort study will be challenging in terms of patient recruitment and retention, as well as sample storage and data management. The potential risks to sample donors will also differ depending on the goals of the biobank and the nature of the specimens collected. For large-scale studies of how the human microbiome evolves over time, many donors will be healthy individuals and specimens collected for these purposes will most likely be fecal specimens or samples from skin, nose, mouth, and vaginal swabs. Because these sample collection techniques are benign, the physical risks introduced during sample collection will be low. By contrast, to study whether changes in the microbiome are predictive of particular illnesses, it will be necessary to collect samples from some individuals with the disease condition. Depending on the nature of the disease, more invasive procedures, such as gastrointestinal biopsies, may be required.

HUMAN MICROBIOME PROJECT BIOBANK INITIATIVES

At present, the Human Microbiome Project (HMP) employs multiple strategies to handle the data and samples that are used in research. The explicit policy of the HMP is that "pre-publication metagenomic and associated data should be released to the scientific community as quickly as possible" (http://commonfund.nih.gov/hmp/datareleaseguidelines.aspx). To this end, the HMP includes a number of member organizations and resource repositories such as the Biodefense and Emerging Infections Research Resources Repository (BEI) and American Type Culture Collection (ATCC). These organizations and resource repositories connect to a centralized data and metadata repository referred to as the Data Analysis and Coordination Center (DACC). The DACC allows for rapid sharing of information among researchers working on projects that may be either intimately related or tangentially connected. Openness characterizes HMP policy to facilitate rapid advance of this area of study.

The HMP's information-sharing requirement has two significant advantages. First, this policy preempts some standard intellectual property concerns.[4] Intellectual property claims typically arise when one group either has exclusive access to a data set or has the exclusive rights to a diagnostic tool. While exclusivity may allow groups working in isolation to profit from what they learn, it slows data and resource sharing by limiting access and use of information and samples. By requiring the early sharing of data and samples through groups and organizations like BEI, ATCC, and DACC, the structure of the HMP obviates these concerns and, in turn, increases the speed with which researchers can build upon the work of others.

Second, this policy answers the question of how widely dispersed and how accessible the samples will be for researchers. When data and samples are made available rapidly and widely, a large number of researchers working on different, perhaps unrelated, avenues of research will have almost immediate access to all microbiome-related samples and research findings. Eliminating obstacles to data sharing increases the diversity and the speed of the uses of samples and data.

As a result of this open sharing policy, substantive confidentiality protections are needed. What these entail depends on the links between samples and data. In cases where data and samples remain linked, the level of oversight and the policies governing the distribution of samples and information should be more restrictive. Steps to limit risks of disclosing confidential information, including prohibitions against comparisons of data in biobank data sets with data available through public sources, will have to be designed and enforced. These are issues that we take up in more detail in the next section.

Ethical, Legal, and Social Implications of Human Microbiome Biobanks

The ethical, legal, and social implications (ELSIs) raised by microbiome biobanks are similar to the many concerns raised by other types of biobanks and the use of stored samples of excess tissues after biopsies or surgery. For example, the risks and benefits of proposed projects have to be assessed and compared to determine whether or not the study is justified. Issues of consent, privacy, and confidentiality are pertinent to discussion of biobanks in general. Nonetheless, because the role that biobanks play in microbiome research is

---

[4] For a more robust discussion of HMP and intellectual property concerns, see Chapter 3: Property and Research on the Human Microbiome.

still unclear, and because this work is still in its infancy and the regulations governing biobanks are currently in flux, these issues demand attention. In what follows, we outline the ways in which the existing federal regulations are an awkward fit for biobank research. The current requirements for informed consent in research cannot be accommodated by biobanks. Moreover, the definitions of "human subject" and "research" introduce ambiguity into determination of whether biobanked samples are subject to these federal regulations.

FEDERAL REGULATIONS APPLIED TO BIOBANKS

As a part of the medical research enterprise that interacts with live human subjects, biobanks are subject to the Code of Federal Regulations 45 Part 46: Protection of Human Subjects (CFR 45 Part 46, often referred to as the Common Rule because it governs research supported by those federal agencies that are signatories to it). Although its origins can be traced back to the Nuremberg Code and the Declaration of Helsinki, CFR 45 Part 46 arose as a result of the 1974 Research Act that produced and empowered the National Commission for the Protection of Human Subjects of Biomedical and Behavioral Research. In 1979, this commission produced the Belmont Report: Ethical Principles and Guidelines for the Protection of Human Subjects of Research (http://www.hhs.gov/ohrp/humansubjects/guidance/belmont.html). This report provided the framework for CFR 45 Part 46.

There are substantial differences between clinical trials and the research that makes use of biobanked samples.[5] The Nuremberg Code, the Declaration of Helsinki, and the Belmont Report were promulgated in response to particular historical research atrocities and designed to address the problems that allowed those unethical uses of human subjects to occur (Beecher 1966). In light of those documents, the Common Rule was designed to protect the National Institutes of Health (NIH) from lawsuits related to the clinical trials it supported (Stark 2012). As a result, the Common Rule is most applicable to clinical trials and studies that impose significant risks to participants. The application of these same regulations to biobank studies restricts research without producing compensating protections for human subjects. This disjunct arises because the regulations were designed to solve particular historical problems related to exposing human subjects to unnecessary or extreme risks of harm.

In some respects, the Common Rule provides inadequate guidance for evaluating research that uses biobanks. In other respects, the existing regulations create unreasonable barriers to socially important biomedical progress.

---

[5] We contrast *clinical research* with research that makes use of biobank samples. Once the sample is taken, future studies typically do not involve doing anything to the body of the sample donor. *Clinical research* makes use of clinical procedures and settings and typically involves doing something to the body of the research participant.

For example, two substantial difficulties in the application of CFR 45 Part 46 to biobanks relate to the regulation's definitions.[6] Namely, the definitions of "human subject" and "research" are incompatible with biobank studies.

According to CFR 45 Part 46, "research" is defined as "a systematic investigation, including research development, testing and evaluation, designed to develop or contribute to generalizable knowledge" (45 CFR 46.102, 2005). The acquisition of samples for biobanks/biorepositories does not fit this definition. When collecting biobank samples, typically there is no systematic investigation in progress. That is, the samples are being collected, but often the disease (let alone the possible diagnostic or intervention responses to a disease) remains unknown. It is also at least possible that samples from a biobank will never be used for any study: they will never be part of a research protocol. Even when biobanked samples are gathered in the process of some other clinical trial or treatment protocol, they will have as-of-yet underdetermined research purposes. Research, per se, as defined in CFR 45 Part 46, does not occur until an investigator formulates a testable hypothesis and begins investigating. All of this means that at the time of donation, when donors contribute their samples and personal information, the activities involved do not meet the definition of research.

Furthermore, to fall under CFR 45 Part 46, studies have to be classified as "human subjects" research. A "human subject" is defined as "a living individual about whom an investigator (whether professional or student) conducting research obtains (1) Data through intervention or interaction with the individual, or (2) Identifiable private information" (45 CFR 46.102, 2005). To be considered "human subject" research, the sample and data must be gathered through intervention or interaction or include identifiable private information.

Of course, the data that biobanks use are gathered through interaction, but at the time when the donor and biobank personnel interact, it is not yet research. It is only later on, when the research protocol is created and initiated and the samples selected for the study, that the sample and data become part of research. At the time that it qualifies as research, though, there is no longer any interaction, so it no longer meets that qualification (#1) for "human subject." Nonetheless, the information used in research may be identifiable and private, indicating that this research could meet that qualification (#2) for "human subject." Whether a future study will actually require any identifiable private information will not be known until long after the interaction with the donor has concluded. As a result, when samples are gathered from a donor, it

[6] This discussion of these problems related to applying the Common Rule to biobanks is not meant to be exhaustive. It is intended to illustrate that even the most basic elements of the regulations (e.g., what counts as consent, what counts as human, and what counts as research) are incommensurate with studies involving biobanks.

will not be known at that time whether the donor is involved in human subject research or not.

Most types of medical research (whether clinical trials of new drugs or medical devices, observational studies of individual behaviors and activities, or epidemiological studies of health outcomes) provide the participants with detailed and specific information about the research project and obtain their informed consent for a narrowly defined research agenda. Study participants are enrolled for discrete periods of time, with the option to withdraw from the study at any time and for any reason. For some biobank donors who are recruited for a specific study, informed consent procedures are similar to those for other types of research. Because the *raison d'être* for most biobanks is to have samples available for multiple potential research projects well into the future, contributors are often asked to provide "blanket" consent for the use of their samples and associated information, often without limitations on the types of research or on sharing specimens and data with other investigators (Murphy et al. 2009). Donors are, however, typically allowed to change their minds and withdraw their sample from a biobank so that it will not be used in any future study.

Under standard expectations of Institutional Review Board (IRB)-reviewed studies, human subject research requires the informed consent of each participant. This includes informing participants about the goals and objectives of the study, how these will be accomplished, and any foreseeable risks. At the time of biobank sample donation, this is not entirely possible. If samples are linked so that donors can be contacted, theoretically, consent would be possible for those donors who are still alive and who can be located. But even for these donors, locating them and obtaining consent would likely be a significant burden to the researcher, particularly when the sample use occurs long after the original sample donation. Furthermore, attempts to contact sample donors to obtain informed consent may involve inadvertent violations of confidentiality, and, depending on the means of contact used, it could certainly create sampling bias.

These terminological and feasibility issues leave researchers in a quandary because biobank research does not meet the strict criteria that allow research to be conducted without informed consent. One possible solution is to declare that biobank research is not human subject research. Another more straightforward solution is to exempt biobank studies from some current informed consent requirements.

One way to avoid the regulatory quandary would be to follow the lead of Vanderbilt University Medical Center and make all samples anonymous. This eliminates the need for informed consent and IRB review and approval of studies that make use of biobanked samples. This choice, however, comes at substantial cost: it severely curtails the usefulness of the samples. For example, any biobank research that would require follow-up with subjects (e.g., updates on medical conditions) could not be done because the researcher could never know the identity of the donor or any characteristic of the donor. This would prevent scientists from pursuing many of the goals of the HMP, such as studies designed to determine whether changes in the microbiome can be used to predict disease onset and severity.

RISKS AND HUMAN MICROBIOME BIOBANKS

Most research involves some degree of physical, social, or psychological risk. Thus, research ethics requires that the risks of harm introduced by research participation be balanced against the anticipated benefits. Physical harms include the pain and bruise of needle pricks and the nausea and anemia that may follow chemotherapy. Limiting participant exposure to risks is accomplished through choice of study procedures or interventions, the selection of the delivery mechanisms, the timing of the interventions, adherence to clinical guidelines, careful oversight, careful recording and review of complications, and ensuring the technical skill of those performing the procedures.

Most human microbiome samples will be gathered through noninvasive or minimally invasive means (e.g., a blood sample or a swab). The physical risks involved in gathering other samples may be somewhat more substantial. For example, the microbiome of the gut is of particular interest. Gathering samples from this microbiome site could require taking an additional small biopsy during a diagnostic procedure or treatment, thereby adding a minimal additional risk to conduct the research. Because these sample donors are already subject to the risks involved as part of their clinical care, contributing a sample to a microbiome biobank would not significantly increase the risk of physical harm. Furthermore, regardless of the type of sample collected, once it is obtained further physical risks of physical harm are nonexistent.

Psychological research risks include reactions like anxiety and depression. For example, some individuals may worry obsessively about what will happen to information they provide to researchers. This risk can be contained by trying to foresee common reactions and by carefully selecting subjects so as to avoid involving those who are most vulnerable to extreme psychological reactions.

Other psychological risks are associated with disclosure of an incidental finding (IF). IFs are discoveries outside of the focus of the study about a research subject's health or disease risk that are noticed in the course of research. As Wolf et al. (2008) have explained, genetic research and research into archived data have produced IFs in the past and continue to do so. Samples and information from microbiome research may also reveal IFs about disease susceptibility. Some researchers and bioethicists are concerned about the possibility that sharing IFs with research subjects could produce psychological distress or depression. Studies are currently under way to ascertain how likely and serious reactions to IF disclosure are, to learn how likely it is that study subjects would want this information, and to assess subject needs and the burdens on researchers associated with providing IF information. As this data is being collected, researchers and bioethicists are also trying to develop guidelines for determining which kinds of IFs should be disclosed (e.g., treatable or actionable conditions, serious conditions, etc.), how they should be disclosed, the resources that should be in place to support the disclosure (e.g., procedures for confirming the IFs, access to genetics counseling), and what study participants should be told about the possibility of IFs when they provide samples to biobanks. Information obtained through a study may also subject the donor to social risks. For example, if a study reveals a medical problem or high risk for a disease (either as an IF or as part of the study protocol), it could make the individual ineligible for health, disability, or life insurance coverage or increase the likelihood that the person will be rejected in an application for employment. As we have seen historically, some medical problems will also carry social stigma and victims may be ostracized. Also, in the current U.S. climate of immigrant demonization and with depictions in the media of how DNA can be used for identification, people could be concerned that their samples might be used to identify them to immigration officials or the police. In such cases they could envision serious social harm in the form of deportation or incarceration. At this time, however, it remains unclear what might be discernible about an individual with only a sample of his microbiome. It also remains unclear how microbiome samples and associated data may be used. As such, the social risks of human microbiome research are not easily predicted.

These social risks could, in turn, exacerbate possible psychological risks. People can legitimately fear discrimination, deportation, incarceration, and associated harms. Extending to microbiome information the existing protections of genetic information with respect to employment and health insurance could go a long way in reducing the social and psychological risks involved in microbiome research. To protect biobanked samples and data from being accessed by legal authorities during criminal or civil proceedings, many biobanks obtain a certificate of confidentiality from the NIH. Some have argued, however, that NIH certificates of confidentiality do not provide absolute protection and are, therefore, inadequate privacy protections (Hermos and Spiro 2009). It will be important to determine whether such certificates of confidentiality are sufficient to protect study participants from social and psychological harms and, if necessary, devise additional mechanisms to protect biobanked samples and data from legal subpoena.

As we have proposed previously (Rhodes et al. 2011), a new conception of risk (*de minimis* risk) should be added as a subcategory of risk for research involving human subjects. Because the risks of biobank and sample bank research are often negligible, they involve only the lowest measure of "minimal risk" as defined in the regulations. Thus, the risks of biobank and sample bank research are more appropriately described as *de minimis*. The role of *de minimis* risk in research is spelled out in more detail in the Chapter 7: Public Health and Research on Populations. It entails a degree of risk so low that harms are nominal and unlikely. A cheek swab is a paradigm example of *de minimis* risk. After obtaining the sample, the risks to human subjects is *de minimis* because nothing further is being done to the body.

BENEFITS AND HUMAN MICROBIOME BIOBANKS

The ethical conduct of research requires that any risks to human subjects be balanced against anticipated benefits to both the individual and society. For the foreseeable future, the benefits of microbiome research will not accrue directly to the individuals who contribute their samples. Nonetheless, the expected societal benefits of advancing this new area of biomedical science justifies the small and unlikely risks to individual subjects, particularly if the safeguards that we recommend for protecting the confidentiality of samples are put in place. As Emanuel, Wendler, and Grady (2000) explain:

> To be ethical, clinical research must be valuable, meaning that it evaluates a diagnostic or therapeutic intervention that could lead to improvements in health or well-being; is a preliminary etiological, pathophysiological, or epidemiological study to develop such an intervention; or tests a hypothesis that can generate important knowledge about structure or function of human biological systems, *even if that knowledge does not have immediate practical ramifications*. (p. 2703, emphasis added)

Rapid advances in technology and the development of new techniques and strategies to isolate and culture the human microbiome make such a promise of societal benefit very likely. Research on the relationship between humans and the microbes that colonize them promises to open doors to understanding human disease in general and may help to reveal critical factors about human disease resistance and susceptibility (see Lee and Mazmania 2010). To the extent that contributing samples for human microbiome biobanks involves only minimal or *de minimis* risk or adds only a minimal risk to a clinically indicated procedure, the risk–benefit ratio argues for gathering and storing samples and data in biobanks.

OBTAINING CONSENT FROM HUMAN MICROBIOME BIOBANK DONORS

The moral and legal obligations to obtain informed consent from human subjects during research are well established. As described by Emanuel et al. (2000), "the purpose of informed consent is 2-fold: to ensure that individuals control whether or not they enroll in clinical research and participate only when the research is consistent with their values, interests, and preferences" (2706). As described previously, however, current Common Rule requirements for obtaining informed consent are not compatible with biobank studies. Consider, for example, the very first sentence listing the regulatory requirements for adequate informed consent. Specifically, informed consent requires: "A statement that [1] the study involves research, [2] an explanation of the purposes of the research and [3] the expected duration of the subject's participation, [4] a description of the procedures to be followed, and [5] identification of any procedures which are experimental" (Code of Federal Regulations Title 45, Part 46.116, 2005). Three of these five regulatory requirements present insurmountable obstacles for biobanks at the time of sample collection. Although [1], a statement that the purpose of the sample collection and retention is to provide material for future research projects, can be provided along with [4], a description of the procedures to be used in obtaining the sample, the specific research protocols that will use biobanked samples in the future are typically not yet developed at the point in time when the subject consents to providing a biobank sample. Thus, it is impossible, at that time, for biobank personnel to provide [2] an explanation of the purposes of the research or [5] identification of experimental procedures.

Finally, requirement [3], which requires researchers to specify the duration of subject participation, is also problematic in the context of biobank research. Does the duration of subject participation include the interaction during which the sample is collected, the length of time during which the contributed sample will be maintained in the biobank, or the period when the particular study using the same is under way? Interactions between sample donors and biobank personnel are typically quite short—the time it takes to gather the sample and demographic information—but a research project can last years and samples can be stored in a biobank indefinitely. Even though the individual donor may not remain in contact with the biobank (she may, in fact, never have more than a single interaction with the biobank), her sample may continue to be used for a length of time that cannot be known in advance.

The mismatches between current regulations for informed consent and the structure of biobanks illustrate how problematic it is to apply these regulations to research using stored samples and data. Nonetheless, participants should be informed about both the possibility of being recontacted and the indeterminate time frame during which the sample may be used. Merely describing this duration of involvement as "indeterminate" is accurate, but it fails to provide a precise picture of the expected duration.

Several methods for eliciting consent from biobank contributors have been discussed in the literature. Specified consent is the closest match to the requirements of the Common Rule and is the standard consent used for clinical trials. In specified consent, consent is given to the specified procedures designated in a research protocol. The most obvious and substantial advantage of specified consent is its provision of comprehensive information. Because the purpose of the study and the procedures and their anticipated risks are laid out in detail, potential contributors can evaluate the burdens and risks to which they would be subjected and endorse the specific research project to which they contribute.

For clinical trials, this kind of specificity about aims, procedures, risks, and burdens tends to be rather straightforward. There is a predetermined hypothesis or issue in question and a protocol designed to produce results with adequate power and confidence. In such cases, a written informed consent form will contain a description of the research protocol, funding sources, a list of procedures and their risks, and the benefits (e.g., contribution to understanding the disease in question) that it is expected to produce upon the successful completion of the trial. Specified informed consent is especially important in clinical trials because these studies often involve burdens and risks that could seem reasonable to some and excessive to others.

Blanket consent is one means to avoid the difficulties associated with obtaining specified consent for research involving biobanks. This is a procedure already used in some existing biobanks. In giving consent to contribute a sample, a donor agrees to have the sample used in any and all future research projects. This approach is appealing precisely because it accommodates the fact that the donor cannot give specified consent at the time of sample contribution because the future uses of the sample are unknowable at the time when the sample is provided.

Although blanket consent avoids the need for recontacting, it also precludes a donor from evaluating the goals and implications of the studies in which the sample is used.[7] That is, individuals who give blanket consent will not be provided with specific and detailed information about what will be done with their sample in the future. Even though biobank studies involve no future physical risks to donors, they will be unable to evaluate the value of the study in light of their own sensibilities. In this way, blanket consent precludes contributors from making personal value judgments about the use of their contributions.[8] For example, some individuals may not want to contribute to certain areas of research (e.g., fertility). To the extent that such sensibilities

---

[7] For a more detailed discussion of these issues see Arnason (2004), Caulfield (2007), and Greely (2007).

[8] Economic and ideological concerns can raise potentially contentious issues related to the indeterminacy of the purpose of the research for which samples will be used. In economic terms, research into the microbiome will be profitable for some companies and the research that makes these profits possible will arise, at least in part, from samples contributed to biobanks. Despite the recent $21 million settlement Dannon made as a result of dubious claims of health benefits associated with probiotics, profitable uses of microbiome research are expected. Accordingly, donors may agree to contribute samples of their gut microbiome to a biobank but would not agree to contribute samples that will be used by a for-profit organization.

In ideological terms, a donor may be willing to contribute samples of her vaginal microbiome for a number of purposes but would be opposed to the use of this sample for the production of new techniques for birth control or abortion.

may be relevant to an individual, blanket consent denies them the opportunity to make informed decisions about the use of their samples in future research projects.

Tiered consent is another alternative that has been discussed by Mello and Wolf (2010), among others. Tiered consent allows donors to provide different levels of permission. At one level, donors may agree to allow their samples to be used in an unrestricted set of future studies. Alternatively, they may limit their permission along several tiers ranging from permission to use a sample for any related study to permission for only a single specified study. They might also request to be recontacted for any studies for which they have not already given specified consent.

Tiered consent allows donors to choose the type of consent they give, but it does so at substantial practical costs. This approach attempts to address some of the practical problems associated with the impossibility of providing the information required for specified consent at the time of sample collection. At the same time, however, it complicates the initial consent procedure because it requires that researchers provide an explanation of all of the relevant tiers in a way that is comprehensible to potential donors. Furthermore, employing tiered consent requires maintenance of detailed records that distinguish the level of consent provided by each donor, mechanisms for recontacting donors who wish to approve each research use of their samples, and resources to address all of these requirements.

Another Consent Option: Process Consent

We recommend a previously unnamed version of consent that we call "process consent." Process consent provides contributors with a description of the process that governs the distribution of samples and associated information to future researchers. Process consent, like blanket consent, avoids the need to recontact contributors and obtain additional consent for each new research protocol, while allowing the use of collected samples and associated information. It requires biobank administrators to identify, in advance, the procedures and criteria by which future research endeavors will be evaluated. These evaluations determine which future projects will be given access to the biobank's samples and data and which projects will not. While typical blanket consent simply gives biobank administrators unlimited discretion in the use of samples and information, process consent restricts the use of samples to projects that are approved by a specified criteria-based review process. In this way, process consent avoids the disadvantages of specified and tiered consent while adding a measure of assurance to blanket consent.

Process consent has several distinct advantages. Process consent relieves contributors of the dilemma of either refusing participation in biobanking or giving unqualified consent to all future research studies. Process consent also allows biobank administrators flexibility in determining the kinds of research projects for which their biobank will provide resources and allows them to limit the personal information linked to the samples provided.

Biobank administrators may take on the role of an "honest broker"[9] or designate an independent group to fill this role. This mechanism establishes a firewall between donor and researcher and determines the appropriate uses of biobank resources according to established and transparent criteria. Incorporating such a mechanism into biobanks provides an independent judgment as to the value of the research project and the need for the requested personal information. It also provides biobanks with the flexibility to provide samples and information to a diverse set of research projects, with various levels of data linkage, data or samples that are identifiable, coded, or anonymized. Confidentiality is thereby ensured to the necessary level by limiting access to identifying information.

International Biobanks

As stated earlier, some microbiome researchers are interested in using biobanked samples to understand whether or not all human beings share a "core" microbiome surrounded by a shell of diversity. Others want to understand how the microbiome is affected by individual genotypes and environmental exposures. These questions can only be answered by sharing data and samples from populations across national boundaries. Thus, international standards governing the collection, storage, and use of microbiome material, particularly samples and data stored in microbiome biobanks, are needed.

Biobanks are being established around the globe, yet data sharing between biobanks varies according to the amount and types of data and samples stored. To the extent that the format, organization, and content of these data sets are not always standardized, integrating information and samples from multiple databanks can involve substantial effort, time, and money (Muilu, Peltonen, and Litton 2007). As Knoppers (2005) noted, early standards for genomic research demonstrated a "failure to anticipate the need for rules for population biobanks and longitudinal studies" (p. 9). Specifically, she notes the abject failure to have a common lexicon for describing different kinds of data. To ensure the usefulness and accessibility of information, common standards for the types of information gathered and the methods for maintaining records should be established. Some models are already in place. The pan-European Biobanking and BioMolecular Resources Infrastructure (BBMRI) project is

---

[9]Recent efforts of the Secretary of Health and Human Services' Advisory Committee on Human Research Protections (SACHRP) regarding the application of the regulations outlined in CFR 46 Part 45 to biobanks also recommended that specimen collection and distribution be governed by "honest broker" mechanisms.

designed to facilitate communication among biobanks and the international researchers who use them (Yuille et al. 2008).

The guidelines for clinical trials outlined by the International Conference on Harmonisation of Technical Requirements for Registration of Pharmaceuticals for Human Use (an ongoing project founded in 1990) should also prove useful in facilitating data sharing. Other international policies may, however, impede collaborative studies. Specifically, in 2003 the United Nations issued an International Declaration on Human Genetic Data, which declares genetic data to be special, and it takes a markedly protective stance toward the use of genetic information. For example, Article 4 of the Declaration states:

> (a) Human genetic data have a special status because:
>
> (i) they can be predictive of genetic predispositions concerning individuals;
>
> (ii) they may have a significant impact on the family, including offspring, extending over generations, and in some instances on the whole group to which the person concerned belongs;
>
> (iii) they may contain information the significance of which is not necessarily known at the time of the collection of the biological samples;
>
> (iv) they may have cultural significance for persons or groups.
>
> (b) Due consideration should be given to the sensitivity of human genetic data and an appropriate level of protection for these data and biological samples should be established. (UNESCO 2003)

This expression of genetic exceptionalism reflects fear and bias on the part of UN policymakers, rather than evidence and reason. It also expresses a social responsibility to protect the public from imaginable harms, regardless of how small or how unlikely these harms may be, particularly for activities like genetic research or sample biobanking.

Genetic exceptionalism turns out to be a real problem for microbiome research. Scientists have recently come to believe that the microbiome samples stored in biobanks and posted on shared websites include human DNA. With current technology, this mixture of DNA may be unavoidable. Even though harms to individuals related to this information sharing are hard to imagine, it is easy to foresee that research regulators will treat this as a serious problem.

Writing in the same vein as the UN Declaration, Bovenberg (2005) suggests that this special status of genomic information implies that the research be governed by public input. But this seemingly benign suggestion is fraught with difficulty. Such governance could involve elicitation of views, or it could involve direct oversight in which the public plays an active role in determining the activities of the biobank. Either route could establish a dangerous precedent of allowing fear, superstition, or political agendas, rather than evidence, to dominate (and obstruct) the pursuit of knowledge.[10] There are other ways of engaging people from the community in research so that their participation can enhance research efforts rather than obstruct them, particularly through the use of community consultation.

Community Consultation: A Reasonable Approach to Participatory Research

Many people in the United States are unfamiliar with the wide variety of purposes for which biobanking is used in biomedical research. Even though the sequencing of the human genome garnered front-page headlines in 2006, genomics is still far from a household word. Genetics research also continues to evoke skepticism and concerns of misuse from the public and members of racial and ethnic groups who have historical reasons for their distrust of medicine and research.

For biobanks to produce meaningful and valid results, participation by individuals from diverse communities and population groups is needed. People from all of our communities must be asked to participate by contributing their DNA, tissue, and microbiome samples. Thus, an informed and willing public is an essential component in achieving the goals of a project like the HMP.

Community consultation and community-based participatory research are models increasingly employed to involve communities and their representatives in conversations about research projects being conducted in their communities. Conversation with members of the communities where the research will be conducted also provides investigators with an opportunity to learn about the values and cultural norms of that specific population so that they may be addressed in the design of the study.

In some models of community consultation, community participants are passive recipients of information from researchers. In others, community participants are engaged in a two-way dialogue. Some biobanks take the model a step further and include community members as active participants in making decisions on the studies that should be allowed to use the biobank.

Community consultation rarely involves asking for community approval, permission, or consent.[11] Consulting with a community ideally includes

[10] John Stuart Mill (1869) has noted the power and undue influence of small and vocal minorities. Even though there are individuals who may object to microbiome research and its application, this is a typical response to innovation and change. While it will be important to sort out just who gets a say and how much of a say, this project is beyond the scope of this chapter.

[11] There are instances when consent will be necessary to engage community members in research. For example, the HapMap Consortium project could not have gotten permission to gather DNA samples from the Yoruba population in Nigeria, Africa, without the consent of tribal chiefs (Rotimi et al. 2007).

eliciting feedback, suggestions, and criticism of a proposed research project. It does not ask for approval or permission from participants because the consultation participants have no legal or moral authority to make decisions on behalf of others in the community.

A community consultation begins by identifying the group most likely to be involved in the study. It then engages representatives of that group in a conversation about the project to elicit the salient aspects of their perspectives and values and to incorporate these into recruitment and enrollment strategies. For example, community consultation for inflammatory bowel disease (IBD)-related studies might involve consulting with IBD advocacy groups, people who have IBD, and potential study participants (Valdiserri, Tama, and Ho 1988).

Selecting the appropriate community consultation model and identifying the relevant population and community are important steps in the process. The concept of "community" is vague, and any individual may identify with a number of different communities. In the context of community consultation, a "population" may be a group of individuals who have a common geographic ancestry (Sharp and Foster 2000). A "community" is a group within a population with many local units of social organization. In one sense, community may be defined by whatever is commonly accepted among a set of individuals; in another sense, it is defined by the objective realities of a biological study. If there is a criterion that delineates a community, it is a set of individuals with some common characteristics—genetic, physical, and social environment and/or geographic. For example, Taiwanese Chinese living in New York City could be considered a community, as might African Americans residing in Harlem, or Hasidic Jews in Boro Park, Brooklyn. At the same time, diabetic patients on Chicago's South Side or individuals with acquired immunodeficiency syndrome (AIDS) from East Los Angeles might also be considered a community, as might members of Alcoholics Anonymous.

Individuals are also not limited to a single community: many individuals consider themselves to be a member of multiple, fluid communities. Even when people identify with a community, they may not want or trust others in their community, or even community leaders, to make decisions on their behalf about research in their community.

Some communities may not even exist prior to a medical research protocol. In the case of the human microbiome biobanks, the most important criterion could be whether or not individuals reside in an institution's catchment area or have a medical record at the institution. For some studies, having certain physical characteristics that are associated with health or illness may define a research community. Again, a critical task in a community consultation is to identify the community that should be consulted. Whereas definitions of community are neither precise nor completely clear, they aim at identifying sets of individuals who may have a common interest in a particular biobank and its uses.

Community consultation serves several purposes for biobanks and the researchers who use them. It can:

- Build and maintain public trust in research and personalized medicine;
- Inform researchers about the values, cultural and social norms, and language/literacy issues of the populations and communities they wish to study;
- Contribute to mutual respect and effective recruitment, enrollment, and retention of participants over time; and
- Inform the public so as to overcome fears of public stigmatization, discrimination, and violations of privacy, particularly among underrepresented and marginalized communities (Burgess, O'Doherty, and Secko 2008; McCarty et al. 2008; Sterling et al. 2006).

A community consultation should begin by explaining the proposed project—the establishment of a biobank, for instance—and its purposes as fully as possible. This allows researchers to describe the possible uses of samples and data and the potential benefits and possible risks associated with such uses. Such disclosures and discussions will face issues of ambiguity similar to those faced by attempts to achieve informed consent. At the time of collection, not all research uses will be known. Nonetheless, such consultations provide opportunities for the community to identify risks that the researchers failed to notice or failed to weigh appropriately in light of community values. Through interaction, researchers and community members can identify potential problems and develop strategies for enhancing benefits and limiting risks.

The NIH first recognized the value of community consultation for genetic research in the late 1990s. In 2000, the first "Community Consultation on the Responsible Collection and Use of Samples for Genetic Research: A Road Map" was convened by the National Institute for General Medicine Studies. Half of the participants in that consultation represented a range of populations and communities, while the other half represented government agencies. After examining views and concerns about population- and community-based genetic research, these groups identified some overriding themes regarding NIH-supported genetic research, such as the need for community consultation at every stage of research and the importance of considering the impact of genetic research on the community alongside individual risks in research.

Two examples of community consultation provide insight into its value to biobank research. The HapMap Consortium (an international group aimed at producing a haplotype map of the human genome) recently sought to understand and address individual and group concerns about biobank participation in three diverse non-U.S. cultures: the Yoruba of Nigeria, the Han Chinese, and the Japanese (Rotimi et al. 2007). Without compromising the HapMap study design, the engagement of community leaders and organizations in bidirectional dialogue resulted in modifications to consent forms as well as adjustments in the practical aspects of recruitment and sample collection. Similarly, the Marshfield Clinic sought to engage representatives from Central Wisconsin communities in dialogue about their understanding of and attitudes about biobank research prior to sample collection (McCarty et al. 2008). This led to the creation of a community advisory board as a mechanism for ongoing community consultation that would work with investigators in addressing ongoing issues. This community advisory board has contributed to the development of methods and guidelines for sharing biobank data and promoting participation in the biobank within their geographic area.

By contrast, the recent case brought by the Havasupai Indians against Arizona State University can be seen as an example of how genetic research can have a negative impact on a community and as an example of the need for effective community consultation. In 1990, university researchers began using genetic studies to examine a wide range of behavioral and other medical disorders. The consent form that the Havasupai contributors signed offered a broad description of a research project with wide-ranging objectives. Those broad objectives were quite different from the description that Havasupai contributors remember. They recall that the study's purpose was to understand whether the high rates of diabetes in their small community had a genetic origin. When members of the Havasupai became aware that the samples were used for other research projects, including some that undermined cherished myths of Havasupai culture and origin and others that they saw as stigmatizing, the Havasupai sued. Arizona State University settled the suit out of court by paying the claimants $700,000.

An effective process of community consultation in designing the parameters of the study might have avoided this dispute. The Havasupai culture includes an account of the tribe's origin in their canyon. Their origin myth was undermined when the genetic studies by the university indicated that the Havasupai share a common lineage with other groups that came across the Bering Strait. Had the study participants understood that their samples could be used to trace ancestral origins or determine other disease associations, they could have voiced their views about limiting the use of their samples or refused to participate. Also, if the investigators had identified this concern, they could have recognized that this community should not be subjects of their study. Collaboration with a community about project design can provide this kind of critical information.

CHALLENGES IN COMMUNITY CONSULTATION FOR BIOBANKING

Community consultation in the design and administration of microbiome biobanks poses some challenges. It is important to identify the relevant community and community representatives, ensure meaningful contributions from a cross-section of the population, and provide access to adequate information about the biobank in a way that is compatible with the health literacy of people within the community. Incorporating community participation and responding to community concerns require skill, patience, and trust in group processes.

Historically, community consultations have resulted in support of biobanks and their goals of improving medical care and health outcomes. Despite this generally positive result, a number of community consultations have highlighted areas of concern (Godard, Marshall, and LaBerge 2007; O'Doherty and Burgess 2009; Pulley, Brace, Bernard et al. 2008). They revealed that participants had a limited understanding of how the DNA samples might be used (Harmon 2010; Kaufman, Murphy, Scott et al. 2008; Pulley, Brace, Bernard et al. 2008) as well as concerns about the confidentiality of the genetic data (Kaufman, Murphy, Scott et al. 2008). Addressing similar issues and concerns regarding microbiome biobanks would be reasonable components of community consultations.

Not all community consultations have reached the same conclusions about which issues are most significant and the appropriate strategies for addressing the issues that arise. For example, a British Columbia Biobank community consultation (Burgess et al. 2008) concluded with unresolved disagreements about donor compensation, the validity of blanket consent, and the necessity of recontacting donors for consent to use DNA samples, but agreement that conflicts of interest[12] should be *avoided*. In contrast, a community consultation in five different U.S. communities, conducted by Johns Hopkins University, concluded that disclosure of conflicts of interest was required.

The different concerns identified by different communities may reflect differences in the populations involved, in the individuals who participated, in the presentations by the consultation leader, and in the models of consultation used. The British Columbia consultation involved a scientist-community dialogue over a number of days with representatives from diverse communities in Western Canada. By contrast, the U.S. consultation involved gathering information in large public meetings at multiple sites across the country.

As microbiome banking and research advance, community consultation will be an important and powerful tool for scientists to use in educating and engaging the participation of diverse populations and communities. Microbiome biobanks have integral value for the future of microbiome research. To ensure adequate understanding of the purpose of microbiome research and its connection to the contributors, community consultation will be needed. Identifying the appropriate community consultation paradigm, however, will

---

[12]The conflicts of interest that provoked worry were the potential conflicts of interest of the researchers who were collecting and storing the samples.

depend in part on the type of microbiome biobank that is being developed as well as the population that will be asked to contribute.

Discrimination and Other Issues

Research into the microbiome carries with it some of the same concerns associated with genetic research. As we have explained, biobanks that store microbiome samples may be linked to personal information. This raises concerns about possible discrimination, stereotyping, and use of information by law enforcement or immigration officials. Similar concerns about the use of genetic information led to the Genetic Information Nondiscrimination Act of 2008 (GINA) in the United States. GINA prohibits "discrimination on the basis of genetic information with respect to health insurance and employment" (Pub. L. No. 110-233, *The United States Statutes at Large.* 122 Stat. 881). Similar protections should be provided for information arising from microbiome samples. People who donate samples should not be subject to social risks because of their willingness to contribute to biomedical research. Furthermore, because scientists need to study a broad sample of the population to develop an understanding of the human microbiome and how it interacts with our genome and the environment, social policies to protect the confidentiality of biomedical research and allay fears of discrimination are needed.

Conclusion

The use of biobanks, together with the technology that now allows data to be shared around the globe, represents dramatic advances in medical research. Most significantly, the use of biobank data shifts the focus of biomedical research from a few individuals with a common illness to understanding the interaction of genes, microbes, and the environment and their contribution to health and disease. This shift in the methods and focus of biomedical research invites a new perspective on the ethical conduct of human subject research and a willingness to shift the focus of regulatory concerns.

The U.S. regulations for the protection of human subjects in medical research were designed for clinical trials with clearly defined aims that may expose a small group of subjects to possibly serious harm. In contrast, biobanking requires the participation of large populations with absent or incomplete knowledge of the research that will ultimately be pursued while often exposing participants to only *de minimis* risks of physical harm. These radical differences now require an adjustment in our regulations to reasonably accommodate the differences between clinical trials and biobank research. Specifically, process consent should be used instead of blanket or specified consent because process consent provides reasonable constraints on the use of biobanked samples without impeding important scientific studies. Finally, community consultation is an important tool identifying the interests of communities and the general public and improving public confidence in biobank research.

Policy Recommendations

- Biobanks for the human microbiome should include an oversight committee that acts as an honest broker—ensuring that only the personal information of contributors that is critical to the research is released and applying explicitly identified criteria for determining appropriate uses of biobanked samples.
- Biobanks should employ a type of process consent. Under this system, contributors provide blanket consent to have their samples and/or information disseminated to investigators only after their proposal and specific information request were reviewed and approved by an oversight committee.
- Human subject research regulations need to be amended so as to take into account the need for studying a segment of the population much larger than previously studied and the fact that the physical risks involved in biobank research are negligible.
- Community consultation should be incorporated into the establishment, organization, and implementation of human microbiome biobanks.

References

Align® Product Information. 2013. http://www.aligngi.com/information-on-Align-probiotic-supplement. Accessed May 10, 2013.

Árnason, Vilhjalmur. 2004. Coding and consent: moral challenges of the database project in Iceland. *Bioethics* 18(1): 27–49.

Beecher, Henry K. 1966. Ethics and clinical research. *New England Journal of Medicine* 274(24): 1354–1360.

Bovenberg, Jasper A. 2005. Towards an international system of ethics and governance of biobanks: a "special status" for genetic data? *Critical Public Health* 15(4): 369–383.

Brandt, Angela M., and Nicole M. Probst-Hensch. 2007. Biobanking for epidemiological research and public health. *Pathobiology* 74(4): 227–238.

Brothers, Kyle B., and Ellen W. Clayton. 2010. "Human non-subjects research": Privacy and compliance. *American Journal of Bioethics* 10(9): 15–17.

Burgess, Michael, Kieran O'Doherty, and David Secko. 2008. Biobanking in British Columbia: discussions of the future of personalized medicine through deliberative public engagement. *Personalized Medicine* 5(3): 285–296.

Caulfield, Timothy. 2007. Biobanks and blanket consent: the proper place of the public good and public perception rationales. *King's Law Journal* 18(2): 209–226.

Elger, Bernice, and Arthur Caplan. 2006. Consent and anonymization in research involving biobanks. *EMBO Reports* 7: 661–666.

Emanuel, Ezekiel J., David Wendler, and Christine Grady. 2000. What makes clinical research ethical? *Journal of the American Medical Association* 283(20): 2701–2711.

Greely, Henry T. 2007. The uneasy ethical and legal underpinnings of large-scale genomic biobanks. *Annual Review of Genomics and Human Genetics* 8: 343–364.

Goldacre, M., L. Kurina, D. Yeates, V. Seagroatt, and L. Gill. 2000. Use of large medical databases to study associations between diseases. *Quarterly Journal of Medicine*, 93: 669–675.

Godard, B., J. Marshall, and C. Laberge. 2007. Community engagement in genetic research: results of the first public consultation for the Quebec CARTaGENE project. *Community Genetics* 10(3): 147–158.

Hermos, John A., and Avron Spiro, III. 2009. Certificates should be retired. *Science* 323 (5919): 1288–1289.

Harmon, Amy. 2010. Havasupai case highlights risks in DNA research. *New York Times*, April 21.

Ioannidis, John P., and Hans-Olov Adami. 2008. Nested randomized trials in large cohorts and biobanks: studying the health effects of lifestyle factors. *Epidemiology* 19(1): 75–82.

Kaufman, D., J. Murphy, J. Scott, and K. Hudson. 2008. Subjects matter: a survey of public opinions about a large genetic cohort study. *Genetics in Medicine* 10(11): 831–839.

Knoppers, Bartha M. 2005. Biobanking: international norms. *Journal of Law, Medicine & Ethics* 33(1): 7–14.

Lee, Yun K., and Sarkis K. Mazmania. 2010. Has the microbiota played a critical role in the evolution of the adaptive immune system? *Science* 330: 1768–1773.

McCarty, Catherine A., Donna Chapman-Stone, Teresa Derfus, Philip F. Giampietro, and Norman Fost, the Marshfield Clinic PMRP Community Advisory Group. 2008. Community consultation and communication for a population-based DNA biobank: the Marshfield Clinic personalized medicine research project. *American Journal of Medical Genetics* 146A(23): 3026–3033.

Mello, Michelle M. and Leslie E. Wolf. 2010 The Havasupai Indian Tribe case–lessons for research involving stored biologic samples. *New England Journal of Medicine* 363: 204–207.

Mill, John Stuart. 1869. On liberty. London: Longman, Roberts & Green. Bartelby: 1999.

Muilu, Juha, Leena Peltonen, and Jan-Eric Litton. 2007. The federated database – a basis for biobank-based post-genome studies, integrating phenome and genome data from 600 000 twin pairs in Europe. *European Journal of Human Genetics* 15(7): 718–723.

Murphy Juli, Joan Scott, David Kaufman, Gail Geller, Lisa LeRoy, and Kathy Hudson. 2009. Public perspectives on informed consent for biobanking. *American Journal of Public Health* 99: 2128–2134.

National Bioethics Advisory Commission (NBAC). 1999. *Research Involving Human Biological Materials: Ethical Issues and Policy Guidance*. Rockville, MD: NBAC.

National Commission for the Protection of Human Subjects of Biomedical and Behavioral Research. 1979. *The Belmont Report: Ethical Principles and Guidelines for the Protection of Human Subjects of Research*. Washington, DC: U.S. Government Printing Office. http://hhs.gov/ohrp/humansubjects/guidance/belmont.html. Accessed July 21, 2011.

O'Doherty, K. and M. Burgess. 2009. Engaging the public on biobanks: outcomes of the BC biobank deliberation. *Public Health Genomics* 12: 203–215.

Payami, Haydeh, Sepideh Zareparsi, Kim R. Montee, Gary J. Sexton, Jeffrey A. Kaye, Thomas D. Bird, Chang-En Yu, Ellen M. Wijsman, Leonard L. Heston, Michael Litt, and Gerard D. Schellenberg. 1996. Gender differences in apolipoprotein E- associated risk for familial Alzheimer disease: a possible clue to the higher incidence of Alzheimer disease in women. *American Journal of Human Genetics* 58: 803–811.

"Protections of Human Subjects: Definitions" Code of Federal Regulations Title 45, Part 46.102. 2005. "Protections of Human Subjects: General Requirements for Informed Consent" Code of Federal Regulations Title 45, Part 46.116. 2005.

Public Responsibility in Medicine and Research (PRIM&R) Human Tissue/Specimen Banking. Working Group. 2007. Report of the Public Responsibility in Medicine and Research (PRIM&R) Human Tissue/Specimen Banking Working Group: Part I Assessment and Recommendations. Boston, MA: PRIM&R.

Pulley, Jill M., Margaret M. Brace, Gordon R. Bernard, and Dan R. Masys. 2008. Attitudes and perceptions of patients towards methods of establishing a DNA biobank. *Cell Tissue Banking* 9:55–65.

Rhodes, Rosamond, Jody Azzouni, Stefan Bernard Baumrin, Keith Benkov, Martin J. Blaser, Barbara Brenner, Joseph W. Dauben, William J. Earle, Lily Frank, Nada Gligorov, Joseph Goldfarb, Kurt Hirschhorn, Rochelle Hirschhorn, Ian Holzman, Debbie Indyk, Ethylin Wang Jabs, Douglas P. Lackey, Daniel A. Moros, Sean Philpott, Matthew E. Rhodes, Lynne D. Richardson, Henry S. Sacks, Abraham Schwab, Rhoda Sperling, Brett Trusko, and Arnulf Zweig. 2011. De minimis risk: a new category of research risk. *American Journal of Bioethics* 11(11): 1–7.

Riegman, P. H. J., W. N. Dinjens, and J. W. Oosterhuis. 2007. Biobanking for interdisciplinary clinical research. *Pathobiology* 74: 239–244.

Riegman, Peter H., and Evert-ben van Veen. 2011. Biobanking residual tissues. *Human Genetics* 130: 357–368.

Rotimi, Charles, Mark Leppert, Ichiro Matsuda, Changqing Zeng, Houcan Ahang, Clement Adebamowo, Ike Ajayu, Tyoin Aniagwu, Missy Dixon, Yoshimitsu Fukushima, Darryl Macer, Patricia Marshall, Chibuzor Nkwodimmah, Andy Peiffer, Charmaine Royal, Eiko Suda, Hui Zhao, Vivian Ota Wang, Jean McEwen, and the International HapMap Consortium. 2007. Community engagement and informed consent in the International HapMap Project. *Community Genetics* 10: 186–198.

Shade, Ashley, and Jo Handelsman. 2012. Beyond the Venn diagram: the hunt for a core microbiome. *Environmental Microbiology* 14(1): 4–12.

Sharp, Richard R., and Morris W. Foster. 2000. Involving study populations in the review of genetic research. *Journal of Law, Medicine & Ethics* 28(1): 41–51.

Stark, Laura. 2012. *Behind Closed Doors: IRBs and the Making of Ethical Research.* Chicago: University of Chicago Press.

Sterling, Rene, Gail E. Henderson, and Giselle Corbie-Smith. 2006. Public willingness to participate in and public opinions about genetic variation research: a review of the literature. *American Journal of Public Health* 96(11): 1971–1978.

The Genetic Information Nondiscrimination Act of 2008. Pub. L. No. 110–233. The United States Statutes at Large. 122 Stat. 881.

Turnbaugh, Peter J., Ruth E. Ley, Micah Hamady, Clair Fraser-Liggett, Rob Knight, and Jeffrey I. Gordon. 2007. The human microbiome project. *Nature* 449: 804–810.

UNESCO. 2003. Statement on human genetic material. http://portal.unesco.org/en/ev.php-URL_ID=17720&URL_DO=DO_TOPIC&URL_SECTION=201.html. Accessed April 28, 2012.

UK Biobank ethics and governance framework version 3.0. 2007. Available: www.ukbiobank.ac.uk/wp-content/uploads/2011/05/EGF20082.pdf. Accessed March 26, 2012

Valdiserri, R. O., G. M. Tama, and M. Ho. 1988. The role of community advisory committees in clinical trials of anti-HIV agents. *IRB* 10: 5–7.

Wolf, Susan M., Frances P. Lawrenz, Charles A. Nelson, Jeffrey P. Kahn, Mildred K. Cho, Ellen Wright Clayton, Joel G. Fletcher, Michael K. Georgieff, Dale Hammerschmidt, Kathy Hudson, Judy Illes, Vivek Kapur, Moira A. Keane, Barbara A. Koenig, Bonnie S. LeRoy, Elizabeth G. McFarland, Jordan Paradise, Lisa S. Parker, Sharon F. Terry, Brian Van Ness, and Benjamin S. Wilfond. 2008. Managing incidental findings in human subjects research: analysis and recommendations. *Journal of Law, Medicine, and Ethics* 36(2): 219–248.

Yuille, Martin, Gert-Jan van Ommen, Christian Bréchot, Anne Cambon-Thomsen, Georges Dagher, Ulf Landegren, Jan-Eric Litton, Markus Pasterk, Leena Peltonen, Mike Taussig, H-Erich Wichmann, and Kurt Zatloukal. 2008. Biobanking for Europe. *Briefings in Bioinformatics* 9: 14–24.

7

Public Health and Research on Populations

Rosamond Rhodes, Stefan Bernard Baumrin, Martin J. Blaser, William J. Earle, Debbie Indyk, Ethylin Wang Jabs, Daniel A. Moros, Lynne D. Richardson, Henry S. Sacks

Microbes and Public Health

Although clinical medicine follows policies that take the population of patients into account, in practice, clinicians primarily focus their attention on one patient at a time. In contrast, public health policies consider the impact of policies on individuals, while public health agents typically focus their attention on the potential benefits and harms for affected populations and communities. As the Institute of Medicine declared in 1988 and reaffirmed in 2003, public health is what we as a society do collectively to assure the conditions in which people can be healthy (Betancourt and King 2003).

Historically, societies have accepted the assurance of the public's health as an ethical duty of medicine and as a statutory duty of government. A few examples from the ancient and modern world will be useful in illustrating the historical role of public health in civilized society, the methods employed by public health officials, the effects of public health interventions, and the extent to which public health measures make a difference in our lives. These examples also demonstrate that long before microbes were identified and their role understood, public health functions included monitoring and responding to microbial communities.

Even in antiquity, grossly unclean water was recognized as dangerous to human health. Rome provided its home counties with an army to protect its security, roads for deployment and commerce, aqueducts (the greatest public works project of the ancient era) to transport clean water from its mountains to its seaside cities, and sewers to move the effluent to the sea. Beyond these basic services Rome provided bread and circuses, that is, elective sustenance and public entertainments for its citizenry. A state may choose to indulge its citizens with sweets and amuse its citizens with galas, festivals, and competitions, or it may not. These are matters of political savvy. Basic services such as roads, security, water, and sewage disposal, however, have long been recognized as essential state obligations.

The provision of potable water and sewage disposal are on this list to illustrate how wide-ranging obligations of the state can be. Public health concerns all and any issues that can affect or threaten the health of the entire community or a segment of the population and, hence, can be state obligations. Humans are constantly using water—to drink, to bathe, for irrigation, and for animal husbandry. Using bad water is likely to cause illness, make food dangerous to eat, and spread disease. Fortunately, in many places potable water is provided by rain, so the state needs only to provide for its proper storage in cisterns and reservoirs. Natural storage in aquifers and wells can also be monitored, and the supply controlled. Obviously, sewage can also be dangerous. Safe sewage disposal requires elaborate works and supervision, and when dangers are identified, they must be eliminated.

Today, with our current understanding of microbial life, it is common for state and local authorities to restrict activities in critical watershed land (Inland Empire Perchlorate Ground Water Plume Assessment Act 2010), to establish requirements for septic systems, and to inspect private property to ensure that human waste is correctly captured and treated. Efforts to provide clean water to the urban populations of the modern industrial state began in the early and mid-nineteenth century with the hygienic and sanitationist movements in England, France, and Germany (Turnock 2007). The development of "vital statistics" and epidemiology provided the data to clearly demonstrate the association of higher mortality rates with urban crowding and filth. Activists argued that government had a responsibility to provide poor urban populations with pure water and other health-promoting measures such as a clean environment and urban parklands.

The following examples further illustrate how microbiome research is a public health issue. William Osler, in his extensive discussion of typhoid in his textbook of medicine, describes an outbreak of cholera in 1885 in Plymouth, Pennsylvania. This example shows how the actions of individuals can affect the health of a community, how public health concerns can justify limiting individual liberty, and how knowledge from bacteriology can be used to design public health policies.

> The town, with a population of 8,000, was in part supplied with drinking water from a reservoir fed by a mountain stream. During January, February, and March, in a cottage by the side of and at a distance of from 60 to 80 feet from this stream, a man was ill with typhoid fever. The attendants were in the habit at night of throwing out the evacuation on the ground toward the stream. During these months the ground was frozen

> and covered with snow. In the latter part of March and early in April, there was considerable rainfall and a thaw, in which a large part of the three months accumulation of discharges was washed into a brook not 60 feet distant.... About the 10th of April cases of typhoid broke out in the town, appearing for a time at a rate of fifty a day. In all, about 1200 people were attacked. An immense majority of all the cases were in the part of the town which received water from the infected reservoir. (Osler 1899, 5)

Discoveries in bacteriology repeatedly led to government regulation of everyday activities in both rural and urban areas. Discoveries about particular enteric diseases, such as endemic typhoid and epidemic cholera, led to the regulation of both the water supply and the disposal of human waste. The development of clear and coherent regulations required the understanding developed from the study of microbes. Without supporting scientific theories and the data from studies, public health interventions are hard to justify. For example, in 1854 John Snow convincingly traced a local outbreak of cholera to the famous Broad Street pump in London through an epidemiologic analysis. He was, however, unable to explain the cause of the outbreak because he lacked the scientific tools to provide microscopic or chemical analysis of the water (Porter 1999).

Scientific tools are essential to provide a basis for public health initiatives, but governmental will and infrastructure are also essential. The following example demonstrates the crucial role of government in implementing effective public health measures. Late in the nineteenth century, seven distinct waves of cholera swept through Europe and North America. The last wave in 1892 was restricted to the area of Hamburg. In *Health, Civilization and the State*, Dorothy Porter describes this famous epidemic:

> The City-state of Hamburg in late nineteenth-century united Germany was a little island of laissez-faire rule in the middle of an autocratic and hugely bureaucratic state...which...resembled eighteenth century England more than nineteenth century Germany. Hamburgers did not believe in government and public intervention.... [C]holera arrived...in late August and...17,000 people caught the disease and over 8500 died.
>
> The wave of cholera which crossed Europe in 1892 had a negligible effect elsewhere. Even in the town which bordered Hamburg, Altona, the impact was small.... The reason was that Hamburg had delayed and squabbled over instituting sand filtration for the water supply, a measure which had been taken by all other major European cities from the early 1870s. (Porter 1999, 95–96)

Ultimately, the urban environment was modified in response to science and advances in microbiology. These were important forces in shaping public policy, civil engineering, the regulation of waste disposal, food processing, food handling, water supplies, and zoning restrictions around reservoirs, rivers, and watersheds.

Much of the discussion of public health in the bioethics literature focuses on the contentious issues of compulsory vaccination, compulsory treatment, and quarantine. Yet, we often lose sight of the fact that most vaccination and treatment are voluntary and, relative to other disease control measures, present relatively few issues. The excessive media emphasis on the controversial issues obscures the fact that there has already been a remarkable consensus on the truly major issues related to waste disposal, pure water, and food safety. The following are some further examples of public health interventions related to food safety and microbes.

Infant mortality was incredibly high in the urban centers of late-nineteenth-century England and the United States. Unlike our present era in which we associate the cold winter months and indoor confinement with an increase in childhood disease, childhood mortality was previously a summer phenomenon, associated primarily with nonepidemic diarrheal disease (Cheney 1984). Ultimately, public health researchers discovered that this disease was related to the bacterial burden of purchased milk.

Much of the scientific public health work involved in discovering the connection between contaminated milk and diarrheal illness was conducted under the direction of William Park of the New York City Board of Health in the tenements during the first decade of the twentieth century. His studies of children under one year of age revealed that winter mortality of 3% ballooned to a summer mortality of 10%. This childhood mortality was almost all due to diarrheal disease. When stratified by feeding with breast milk or purchased milk and by the amount of bacteria present in the milk supply, those with a bacterially better milk supply had a mortality rate of 2% and those with a poorer milk supply had a mortality rate of 15% (Oliver 1956, chap. 8). This data provided the evidence to justify imposing regulations on the milk supply.

On the basis of a growing body of scientific literature that documented the dangers of other milk supplies, in 1912 New York City enacted an ordinance requiring the pasteurization of all milk sold commercially within the city. This regulation replaced a slightly older and labor-intensive system of inspection and certification that managed to certify only 1% of the milk sold in New York City (Dupuis 2002). Similar regulations were soon developed throughout the country.

As is typical with public health measures, these regulations inflicted hardships on some people but also benefited many more. Pasteurization legislation affected the livelihood of millions of food producers at a time when close to 50% of the population still worked on farms. This was an era when there were still small dairy farms within New York City as well as in its nearby rural suburbs. A report of the New York City Board of Health for 1910–1911 describes a great army of workers, perhaps 300,000, engaged in producing and handling the milk supply (Oliver 1956, p. 292). E. Melanie Dupuis explains the consequences of the pasteurization regulations.

> Pasteurization mandates in cities led to significant shakeouts in the number of firms supplying those cities with milk. For example, in the Milwaukee milk market, the number of milk distributors decreased from two hundred to thirty two between 1914 (when the city pasteurization ordinance went into effect) and 1920.... [T]he implementation of pasteurization regulations transformed an industry of small scale dealers into one of major, large-scale, capital-intensive distributors.
>
> Not long after this shakeout, the dairy industry began a period of intensive merger activity, creating two major dairy companies that dominated many of the eastern markets: Borden's and National Dairy. Beginning in the 1920's these companies initiated expansion policies involving the purchase of hundreds of smaller companies.... By 1932 these two companies purchased one third of all milk in the North Atlantic states and over 20% of all milk in the United States as a whole. (Dupuis 2002, 82)

We now appreciate that all animals have an individual and, to some extent, species-specific microbiome. We do not yet fully understand how unique (as opposed to group or species specific) an individual's microbiome is at any point in time, how much it varies with time, and the extent of its overlap with that of other individuals. To develop that understanding will require an examination of the microbiomes of different human beings as well as those of other species.

The increasing incidence of salmonella infections in the United States is a telling example of how the microbiome in different species can be affected by tampering. According to *Beasts of the Earth: Animals, Humans and Disease* by E. Fuller Torrey and Robert H. Yolken, the modern poultry business involves mass processing of eggs and chicken parts in large automated facilities, increasing the risk of cross-contamination.

> In the 1970's, in an attempt to increase egg production, efforts were undertaken to eradicate two serotypes of salmonella, *Salmonella gallinarum* and *Salmonella pullorum*, that affect chickens and other poultry. These serotypes do not affect humans but cause diarrhea and other illnesses in poultry, leading to decreased egg production. In severe epidemics, these bacteria have been known to kill entire flocks of chickens. *Salmonella enteritidis* is closely related to these two serotypes, and in fact, studies suggest that all three serotypes were derived from a common ancestor. As *Salmonella gallinarum* and *Salmonella pullorum*, which do not cause human disease, were eradicated from chickens, *Salmonella enteritidis*, which does cause human disease, took their place. Thus a flock of chickens may be infected with this bacteria, with no outward sign to indicate potential problems for humans.

> Chickens infected with *Salmonella enteritidis* may transmit infection to humans through their eggs or their meat. You can have an infected but completely healthy looking chicken that lays infected but completely normal looking eggs. (Torrey and Yolken 2005, 103–104)

Torrey and Yolken are describing an unintended consequence of a poultry industry decision to increase efficiency and profit by creating chickens with a variation in their microbiome so as to eliminate a threat to chickens. Unknowingly, it introduced a risk to human consumers. This example is not presented to criticize an industry's effort to improve efficiency and return on investment, or to suggest turning back the clock. Rather, we present this case as a clear example of the complex interaction among the microbiome of different species and how efforts to produce benefits may have unintended consequences. Again, at every point the growth of knowledge has the potential for disadvantaging some and benefiting others.[1] The intended purpose of public health activity, whether the activity is information gathering, research, or enforcement of regulations, is to significantly promote and protect the health of the population.

Public Health, Liberty, and Privacy

In the late nineteenth century, developing knowledge about the spread of enteric disease created a deep conflict between highly valued individual freedoms and property rights on the one hand, and the duties of local and central government to act in the interest of the greater community on the other. The authority of government to act on behalf of public health had long been recognized as part of the state's legitimate police powers, but the appropriate limits of government action were controversial. Controversies over government-imposed health measures therefore arose in response to immediate confrontations with epidemic disease. In controlling the spread of disease, sometimes people's freedoms are limited, and sometimes their property is destroyed. Again, for the benefit of some, others may have to pay a heavy price.

---

[1]Imagine that for public health reasons the microbiome of the chicken were altered back to a point in the past where there was less potential for human disease (a general benefit to consumers of poultry and eggs), but a loss of efficiency within the poultry industry (a potential loss to poultry producers and processors). If the increased cost of producing poultry products had no effect on consumption, then possibly all the increased costs would be passed on to consumers and the poultry industry would suffer little loss of income. Consumers would spend more but, presumably, not so much more as to outweigh the public health gain. If the increased costs to the poultry industry made an alternative food more desirable because of a relatively better price, then the public health benefit would be accompanied by a negative effect on owners and workers in the poultry industry. In either case, the example illustrates how increased knowledge about the microbiome could potentially have a profound effect on people's livelihoods and incomes.

In the late nineteenth century, scientific progress itself also provoked debates. Some people denied the validity of science, and some refused, as a matter of personal freedom of conscience, to accept its conclusions. Certainly, responding to science-based demands can be costly. Even today we are confronted by the science-based demand to reduce our carbon footprints so as to diminish global warming. When the response involves personal sacrifice, self-interest can cloud people's judgment as to whether the burdens of public health interventions outweigh the benefits. Increasingly, in recent decades, public health goals have also been pitted against claims of privacy. In particular, public health activities that involve contact tracing for sexually transmitted infections and questionnaires about personal behavior have raised privacy concerns.

The clash between public safety and health goals on the one hand and issues of liberty and privacy on the other is another important subject of philosophical reflection and controversy. These issues are discussed further in the **Coda** of this chapter.

Public Health Functions

Over the past century, public health systems have developed an expanding array of core functions. As of 1920, the list of public health responsibilities included sanitation, infection control, education in principles of personal hygiene, early diagnosis and preventive treatment of disease, and development of the social conditions necessary for the maintenance of health (Winslow 1920). By 2003, the list had expanded to include assessment of health status and health needs, identification of health threats and risks, policy development and planning to maximize health and minimize risks, assurance that people receive necessary services, and assurance that conditions necessary for health are maintained. Although the extent to which the government has taken adequate measures to achieve these core functions waxes and wanes with politics and the availability of public funding, the list of responsibilities remains as a statement of reasonable expectations.

The human microbiome is likely to play a role in disease susceptibility and resistance, and, to some extent, the bacteria, viruses, and fungi that comprise it can be transmitted through the environment and from one individual to another. Therefore, we can expect that monitoring the human microbiome and responding to changes in it will be a matter for public health involvement in the future. Each of the core public health functions will be relevant to the microbiome. Because disease states can be associated with particular features of a microbiome, we want public health agencies to be involved in microbiome research and policy to protect the health of the community. As we learn more about what constitutes a healthy microbiome, we want public health agents to characterize and monitor the health of the microbiomes within our communities. It is also likely that increased knowledge of the human microbiome will lead to changes in our oversight of sanitation, the water supply, and the environment. Learning about the microbiome will lead to new methods for the assessment of the health status of the microbiome of individuals and communities. In that light, public health agents will be called upon to identify microbiome-related health needs and health threats and develop infection control measures in response. Their involvement will require policy development and planning along with both public education and education of health professionals. This will enable public health agents to develop and employ new tools for early diagnosis and preventive treatment of disease. To assure justice as well as health promotion, public health officials will have to develop means for assuring that the conditions and services necessary for maintaining a healthy microbiome are broadly distributed across the population.

Public Health Agency Powers

Society grants public health agencies broad and effective police powers to enable them to accomplish their core missions. Public health authorities can mandate reporting of disease, and under certain circumstances, they can mandate treatment of diseased individuals. Public health authorities can require everyone to comply with environmental regulations, immunization requirements for humans and animals, disease prevention strategies (e.g., fluoridation of water), and bans on unhealthy behaviors (e.g., smoking in public transportation, buildings, and parks). They can regulate the use and distribution of hazardous materials. They even have the authority to quarantine individuals.

Public health strategies often involve sacrifices of liberty or privacy. Requiring individuals to accept invasions of their bodies (e.g., to be vaccinated), to do things that they may not want to do (e.g., to stay quarantined at home), or to refrain from doing things that they want to do (e.g., using controlled drugs) infringes on their liberty. Disease surveillance also interferes with privacy. Because we cherish our liberty and privacy, these sacrifices must always be justified. Liberty- and privacy-limiting public health measures are justified by the benefits they provide and the harms they avert. In this way, adopting public health policies always involves a risk–benefit assessment and an evaluation of the burdens and benefits imposed on the affected populations.

Appropriate public health policy involves a complex assessment of risks, harms, and benefits that affect the entire population. It involves viewing disease risk as a continuum, not a dichotomy. Some public health interventions involve only a small fraction of the population at high risk for a serious harm (e.g., workers exposed to toxic substances such as asbestos or uranium). Other public health interventions involve exposure of a large portion of the population to the risk of a small and short-lived harm (e.g., pesticide spraying for mosquito control). Other interventions provide a significant benefit to the vast majority while exposing a few people to risk of serious harm (e.g., vaccination) (Rose 1981). On a public health level, an individual's risk of illness cannot be isolated from the disease risk for the population to which she belongs (Battin et al. 2009; Rose 1992). Many effective public health strategies focus on modification of risk for the entire population. Such strategies often require the participation of every individual (e.g., water fluoridation) to provide large benefits to the community while subjecting each participant to some theoretical risk (Rose 1981).

Public Health Methods

The risk and benefit assessments that ideally support public health policy and intervention should be based on systematically collected measurements or data. The data supporting public health policies come from studies that are classified sometimes as surveillance and sometimes as research. Surveillance involves the systematic collection, analysis, and interpretation of outcome-specific data. Surveillance is often an ongoing data collection activity that is coupled with efficient communication of information to public health officials who are responsible for controlling disease and preserving health (Thacker and Berkelman 1988).

Environmental surveillance involves using scientific procedures for measuring sanitary conditions, water supply, food supply, air quality, and environmental toxins. Epidemiologic surveillance entails collecting data from populations. It traditionally focuses on collecting data on communicable and chronic diseases by gathering information from physician and laboratory reports, syndrome surveillance, hospital emergency departments, emergency medical services (EMS), and so forth. Data from surveillance are systematically analyzed by applying scientific and statistical procedures for comparing these measurements and drawing informed conclusions (Fink 1993; Rossi and Freeman 1993).

Public health research traditionally focuses on answering questions about the transmission, containment, and prevention of infectious disease. It often involves surveillance of a population to gather epidemiologic information on chronic and acute illness as well as on environmental factors. This information is useful in identifying factors that may contribute to disease and documenting the magnitude of a problem. Analysis of the data can alert officials about emerging factors that affect health. Public health research also involves evaluation of the need for and effectiveness of public health policies as well as planning for emergency responses to health-related disasters (Langmuir 1980). In sum, it is hard to distinguish public health surveillance from public health research because both activities involve scientific, hypothesis-driven, systematic collection of data with the goal of guiding effective policy.

Reliable public health surveillance and research require very broad public participation because any opting out introduces biases and distorts the picture that the data present. When affected groups are not included, a public health concern may be minimized, and when unaffected groups are not included, a public health concern may be exaggerated. In other words, without broad public participation, selection bias occurs. Using data that are unrepresentative in any way can lead to public health policies that are poorly suited to the situation.

As we learn more about the role of the human microbiome in acute and chronic disease, it is likely to become an increasingly important feature of public health research. Researchers will want to learn about the effect of the microbiome, and changes in the microbiome, on all sorts of disease conditions. They will also want to track changes in the microbiome across the population to learn about the health impact on individuals and particularly on individuals who spend all or most of their time in the same contained environment (e.g., prisons, schools, nursing homes). Similarly, with increasing use of probiotics, public health officials should collect data to assess their impact on the community's health. With the data in hand, public health officials will be able to perform comprehensive analyses to direct and evaluate relevant public health policies and build response capacities for the future.

Research Regulations and Public Health Data Gathering

Public health surveillance and research studies are based on scientific principles and employ scientific methodology. Public health studies, like clinical research studies, start with a theoretically grounded hypothesis about the relationship of some factor to a biological function. These studies then proceed by systematically obtaining data and analyzing it. That data then allows researchers to make general statements about the studied relationships. In other words, both public health studies that are classified as research and studies that are not counted as research but considered surveillance and exempt from the regulations, produce generalizable knowledge.

Today's research regulations were devised with a focus on clinical research, that is, the study of the safety and efficacy of interventions (e.g., drugs, devices, procedures) as treatment for diseases. They were not focused on addressing public health activities, so their application to public health is not always a perfect fit. Nevertheless, the same set of regulations is applied to both domains. Regulatory requirements that express appropriate caution and concern (for example, in clinical trials of new drugs to combat cancer), often seem like unreasonable or insurmountable burdens in public health studies.

The need for surveillance and research as a basis for establishing public health interventions and policies raises its own important definitional and ethical questions. Is public health surveillance research? If it is, does it fall under the regulations governing human subject research?

. The U.S. regulatory distinction between treatment and research derives from the 1979 Belmont Report (National Commission 1979). The definitions of "research" and "human subject" were then articulated in the 1991 Federal Policy for the Protection of Human Research Subjects, under U.S. Department of Health and Human Services regulations at 45 C.F.R., part 46, subpart A, informally known as the Common Rule (Code of Federal Regulations. 1991). According to this stipulation, "research" is "a systematic investigation, including research development, testing and evaluation, designed to develop or contribute to generalizable knowledge".[2] It defines a "human subject" as a living individual about whom an investigator (someone conducting research) obtains (1) data through intervention or interaction with the individual or (2) individually identifiable health information. According to the regulations, all research involving human subjects must be reviewed and approved by an institutional review board (IRB), but there is often confusion or disagreement about which projects count as research and require such review and which count as surveillance and do not.

According to current regulations, public health surveillance involves observational study, and it is exempt from the regulations governing research. Public health surveillance activities may identify pathologic organisms, identify susceptible individuals, and monitor interventions to eliminate dangerous organisms, all without having to obtain individual informed consent required for research.

While important social benefits can be provided by conducting public health research as well as public health surveillance studies, the criteria for drawing a line between what counts as research and what does not are vague. Often there is no discernible difference between scientific activities called "research" and those that are not. Rather, there is only a stipulation framed in uninformative terms. In contrast, the most ethically significant differences between studies are the level of risk to which human subjects are exposed and the importance and likelihood of the expected benefits.

Quality Assurance, Quality Improvement, and Surveillance

Similar issues arise with quality assurance (QA) and quality improvement (QI) activities. QA programs systematically monitor and evaluate various aspects of

---

[2] The same definition of research is also used in the Health Insurance Portability and Accountability Act (HIPAA) Privacy Rule.

a project, service, or facility to ensure that standards of quality are being met. QA data from the systematic documentation and analysis of hospital activities provide the knowledge that guides hospitals to assure the quality of their care. The knowledge generated from QA data allows institutions to detect and rectify practices that put people at risk, including deviations from standard operating procedures and unforeseen results of new cost containment measures.

QI is a formal approach to the analysis and systematic improvement of performance. Data from QI activities lead to knowledge that is used to alter practice to achieve the highest-quality care at the most reasonable cost. Analysis of QI data provides the knowledge base that allows institutions to compare the quality and efficiency of their performance to that of national databases. It also guides them to implement programs that improve patient outcomes and avoid errors (Cohen et al. 2008).

In hospitals and health facilities there is a growing emphasis on quality of care. Thus, hospitals, medical groups, and other health care entities have developed elaborate QA and QI programs to further their health-related goals. These all involve monitoring the degree to which health care systems, services, and supplies are in line with current professional knowledge, and whether institutional practices increase the likelihood for positive or negative health outcomes for individuals and populations. These observational data collection activities may or may not be subject to the regulations governing research. These activities are important as public health measures, but it is unclear whether human subject research regulations apply to them. Some examples will be useful in illustrating the related confusion and controversy.

The Pittsburgh ESRD Case: In October 2000, a physician who was a specialist in kidney disease coauthored an article about a project to improve the dialysis care delivered to patients in Medicare's End Stage Renal Disease (ESRD) program. Sometime after the article appeared, his university notified him that an audit of faculty publications had identified his project as a QI effort that met the definition of human subject research. It had not undergone IRB review.

The nephrologist responded that he had participated in the project as the chair of the local ESRD Network's medical review board, that the ESRD Network conducted the project under contract to the Centers for Medicare and Medicaid Services (CMS), and that the CMS scientific officer overseeing the project had told him that, as a CMS-directed QI project, it was not subject to oversight by the university's IRB.

Later, CMS reaffirmed this opinion in a letter to the university. The university submitted the dispute to the Department of Health and Human Services' Office for Human Research Protections (OHRP) for review. OHRP responded that the design of the quality improvement project met the definition of human subject research and that the CMS determination that the project was exempt from IRB review "was not made in collaboration with OHRP." OHRP and

CMS have had discussions and exchanged memoranda as to whether this and other quality improvement activities meet the regulatory definition of human subject research and require IRB review. As of this writing, however, these issues have not yet been resolved (Baily et al. 2006).

The Checklist Study: Another example of conflicts over interpretations of the regulations is the Checklist Study. This study, led by Dr. Peter J. Pronovost of Johns Hopkins University, was published in the *New England Journal of Medicine* in 2006 (Pronovost et al. 2006). Catheter-related bloodstream infections occurring in the intensive care unit (ICU) are common, costly, and potentially lethal. Dr Pronovost and colleagues conducted a collaborative cohort study, predominantly in ICUs in Michigan. The study attempted to increase clinicians' use of five procedures recommended by the Centers for Disease Control and Prevention (CDC) and identified as having the greatest effect on the rate of catheter-related bloodstream infection and the lowest barriers to implementation. The recommended procedures are hand washing, using full-barrier precautions during the insertion of central venous catheters, cleaning the skin with the antiseptic chlorhexidine, avoiding the femoral (thigh) site if possible, and removing unnecessary catheters. Clinicians were educated about practices to control infection and harm resulting from catheter-related bloodstream infections; a central-line cart with necessary supplies was created; a checklist was used to ensure adherence to infection control practices; providers were stopped (in nonemergency situations) when these practices were not being followed; the removal of catheters was discussed at daily rounds; and the teams received feedback regarding the number and rates of catheter-related bloodstream infections in their unit. Investigators measured rates of infection at 3-month intervals and found that the median rate of catheter-related bloodstream infection per 1000 catheter-days decreased from 2.7 infections at baseline to 0 infections at 3 months after implementation of the study intervention, and the mean rate per 1000 catheter-days decreased from 7.7 at baseline to 1.4 at 16 to 18 months of follow-up. The investigators concluded that use of the checklist and other procedures resulted in a large and sustained reduction (up to 66%) in rates of catheter-related bloodstream infection that was maintained throughout the 18-month study period.

That the use of these simple QI procedures could have such a large benefit was an important finding and was hailed as "remarkable" in an editorial accompanying publication of the study (Wenzel and Edmond 2006). Shortly after the study results were published, however, OHRP wrote a letter to Johns Hopkins University (JHU). OHRP maintained that the implementation of the Comprehensive Unit-based Safety Program that included the ICU Safety Reporting System at Michigan and Rhode Island hospitals, the subsequent collection and analysis of data from ICU patients exposed to those interventions, and the surveys of hospital personnel represented nonexempt human subject research that was conducted without appropriate IRB review and approval, in contravention of U.S. Department of Health and Human Services (HHS) regulations[3] (Pritchard 2008).

In addition, OHRP found that JHU had failed to ensure that all collaborating institutions engaged in the research operated under an appropriate OHRP-approved assurance of compliance. Such an assurance was required by terms of the JHU institutions' federal-wide assurance (FWA)—the document that institutions receiving federal research funding file, promising that all research conducted will comply with federal regulations (OHRP 2007). In other words, OHRP concluded that the study had failed to obtain the necessary IRB approvals before commencing and ordered JHU to suspend the program.

Many took this to be a new and unwarranted interpretation of the regulations, and OHRP was widely criticized. In an op-ed piece published in the *New York Times*, Atul Gawande, a Harvard surgeon who writes frequently about health care for *The New Yorker* magazine, declared that:

> The government's decision was bizarre and dangerous.... Scientific research regulations had previously exempted efforts to improve medical quality and public health because they hadn't been scientific. Now that the work is becoming more systematic (and effective), the authorities have stepped in. And they're in danger of putting ethics bureaucracy in the way of actual ethical medical care. The agency should allow this research to continue unencumbered. If it won't, then Congress will have to. (Gawande 2007)

In an attempt to clarify its position, OHRP then stated that the regulations do not apply when institutions are only implementing practices to improve the quality of care. At the same time, if institutions are planning research activities examining the effectiveness of interventions to improve the quality of care, then the regulatory protections are important to protect the rights and welfare of human research subjects (OHRP 2008).

In other words, according to OHRP, it was permissible to conduct the program only if the investigators were not trying to learn from it. JHU responded with an explanation of their reasoning as well as a number of changes in their procedures and, a few months later, the program was allowed to resume. OHRP explained:

> Based on this description of the Johns Hopkins University projects, our analysis of how the regulations for the protection of human subjects in research (45 CFR part 46) apply to these projects is provided below.
>
> First, we believe that the actual implementation of the five-part catheter-related bloodstream infection reduction program in the participating hospitals is a quality improvement activity that does not meet the regulatory definition of *research*. This is because none of the parties involved are

[3] HHS regulations at 45 CFR 46.103(b) and 109(a).

> implementing the program as a research intervention in order to evaluate its effectiveness. Here, the program is being implemented solely for the purpose of improving the quality of care. (Pritchard 2008)

In other words, it was permissible to resume the program because the investigators were no longer learning from it.

Again, the line between activities exempt from regulation and research activities is often hard to discern. It is not surprising that institutions and public health agencies find themselves in quandaries about how to classify their studies (Kofke and Rie 2003; Lynn et al. 2007). The controversy over whether Peter J. Pronovost's Checklist Study should have been counted as research or QI illustrates the problem. The point is that experts in research ethics disagree about whether the Pronovost study is research or QI (Baily 2008; Flanagan, Philpott, and Strosberg 2011; Gawande 2007; Kass et al. 2003; Miller and Emanuel 2008; Pritchard 2008). There is no ethically or scientifically significant distinction between research and QI studies in that both involve a theory-driven scientific hypothesis, systematic collection of data, analysis of the data, and drawing of actionable conclusions that count as knowledge.

Public health agencies have maintained that the primary purpose of their surveillance activities is to allow them to act in ways that promote health and prevent disease, **not** to produce generalizable knowledge. Thus, they maintain that surveillance does not fall under the research regulations. This response is problematic for two reasons. All knowledge, including knowledge from surveillance, is generalizable in that it can be applied to additional instances. The example presented earlier in this chapter of knowledge about disease transmitted by milk in one city being applied to public health policy in other cities is a case in point. Furthermore, knowledge is the basis for advancing the promotion of health and responding to new or evolving problems. The production of generalizable knowledge is beneficial to the public and should be encouraged, not discouraged. Sorting out whether a public health activity should or should not be considered research does not reflect any legitimate concerns. As Amy L. Fairchild and Ronald Bayer explain:

> The recent efforts to provide definitional solutions to the question of research and public health practice involve twists and turns that inevitably produce results that are riddled with inconsistencies and that are conceptually unsatisfying whether Duncan, S. C. 2010 such activities fall neatly under the classification of research or practice or exist in a gray borderland. (Fairchild and Bayer 2004, 631–632)

To that end, Fairchild and Bayer propose a form of ethical review to balance the tension between the claims of individuals and those of the common good.

The point of Fairchild and Bayer's criticism is that research regulators have relied on their own stipulative definitions of "research" and "human subject" to do the conceptual work of determining the requirements for the ethical conduct of research. In fact, the arbitrary regulatory distinction between public health surveillance, QI, QA, and "research" in the realm of population studies turns on a single word, *designed*. The emphasis on *designed* leads many policymakers and regulators to focus on the intention of researchers in sorting out whether informed consent and IRB review are required.

Defenders of the regulations see a significant difference between *designed* and *intended* that others, including some government agencies, are unable to discern. Published documents provide evidence that people understand that *intention* is the central focus of the regulations. For example, a document on the official CDC website, "Guidelines for Defining Public Health Research and Public Health Non-Research" (October 4, 1999), identifies *designed* as the key term that must be examined in sorting out whether or not a study is to be considered research. There it is explained that:

> The major difference between research and non-research lies in the primary **intent** of the activity. The primary **intent** of research is to generate or contribute to generalizable knowledge. The primary **intent** of non-research in public health is to prevent or control disease or injury and improve health, or to improve a public health program or service. Knowledge may be gained in any public health endeavor designed to prevent disease or injury or improve a program or service. In some cases, that knowledge may be generalizable, but the primary **intention** of the endeavor is to benefit clients participating in a public health program or a population by controlling a health problem in the population from which the information is gathered. (CDC 1999, emphasis added)

A similar document, "Public Health Practice vs. Research: A Report for Public Health Practitioners," prepared by the public health epidemiology organization, the Council of State and Territorial Epidemiologists, offers a similar explanation of this confusing and controversial point. It states that:

> The study's hypothesis, methods, implementation, and underlying **intent** may support a conclusion that the activity is research. As a result, the public health agency must adhere to a series of protections and procedures pursuant to the Common Rule. These protections (including individual informed consent absent a waiver) and procedures (including review by an IRB) are designed to protect the health and safety of human subjects. (Hodge and Gostin 2004, 10, emphasis added)

Many moral theorists, including consequentialists and numerous authors who write about moral responsibility, hold that intention alone does not determine the rightness or wrongness of actions (e.g., Foot 1967; Ross [1930] 2002). As they see it, in our everyday lives we hold people responsible for what they consciously do. To some extent, even when people do not explicitly intend to bring about the consequences, they may be held responsible for them. The person who intends to toast a marshmallow but sets off a massive brush fire is held responsible for the consequences of his carelessness. When BP executives intended to extract oil from their Deep Water rig and make a handsome profit, we held them responsible for the consequences of the oil that escaped into the Gulf of Mexico. In sum, the rightness or wrongness of an action primarily derives from its context and the various factors and justifying reasons that tend to make an action right or wrong (Scanlon 2008). Although philosophers offer somewhat different analyses of what contributes to the rightness or wrongness of action, the foreseeable and likely benefits and burdens that are consequences of the action are often key features in determining whether the action was right or wrong.

The intention of the agent figures primarily in our assessment of the agent's character, less so in the determination of the rightness or wrongness of the agent's action. Someone who wrongs another deliberately is often considered more blameworthy than someone who harms another inadvertently. We are more likely to forgive or excuse a distracted or careless person, who neither intended nor foresaw the consequences of the action, than we are a person who pointedly tried to inflict harm.

Moral judgment typically turns on harms and benefits, reasons, principles, and commitments. It is therefore peculiar, but understandable, that research ethics hinges on the 1974 definition of research. Because of the misplaced emphasis on intention, inadequate attention is focused on weighing the foreseeable risks and minimizing harms to research participants.[4] Whereas the Belmont Report used the term *designed* to stipulate a distinction between public health surveillance (National Commission 1979), QI, and QA on the one hand, and research on the other, other factors are more significant in determining whether a research project should or should not be allowed and what may be done, to whom, how, and when.

De Minimis Risk: A Proposal for a New Category of Research Risk

Critics have noted that current research regulations and IRB policies impede research and limit or even discourage learning from clinical practice (Emanuel and Menikoff 2011; Emanuel et al. 2004; Fost and Levine 2007; Resnik, Sharp and Zeldin 2005; Schwab 2010). Regulations and policies related to informed

---

[4] Aside from the egregious cases where researchers abused individuals from undervalued populations (e.g., retarded children at Willowbrook, poor sick black men in Macon, Georgia, in the Tuskegee study) (Seto 2001), researchers typically consider the risks of inflicting harm and they try to design their studies so as to minimize them. Generally, they have also historically limited the riskiest studies to the sickest subjects, that is, to those who had the most to gain if the experimental intervention was successful and the least to lose if it was not.

consent and privacy protection are especially burdensome, time consuming, and costly (Infectious Diseases Society of America 2009; Kulynych and Korn 2002; Vates et al. 2005). For population studies that pose only negligible risks to participants, it is essentially impossible to meet existing regulatory or institutional requirements, and consequently, many legitimate research projects cannot be undertaken (O'Herrin et al. 2004; Shen et al. 2006). In response to these problems, rather than simply offering modifications to the existing framework, we identify a critical oversight in the regulations and offer a solution that could make them more coherent.

Looking through the lens of microbiome research suggests that we need a fresh perspective and new distinctions for evaluating the ethical conduct of studies on human subjects. Regardless of how or why intention became the lynchpin criterion for conducting scientific studies that bear on public health (Boyle 1980; McIntyre 2009), the need for research on the human microbiome provides an opportunity for taking a fresh look at the ethics of population studies.

In many circumstances, a study may serve a significant public good, and obtaining informed consent may not be possible. For example, the Pronovost Checklist Study could not have obtained consent from many of the ICU patients as they were either unconscious or lacked capacity as a result of their condition. In other circumstances, obtaining informed consent is inefficient and counterproductive. In public health surveillance, for instance, the risks to participants are typically miniscule, and data may be needed from every individual with a particular condition who presents in an emergency department or doctor's office. Obtaining informed consent from each individual would require tremendous effort and thereby divert scarce public health resources from other socially important activities. Again, because obtaining informed consent allows people to refuse or withdraw participation, the results of the study may be biased. Selection bias remains one of the most frequent deficiencies of biomedical studies (Feinstein and Horwitz 1978; Ferreira-Gonzalez 2009; Heckman 1979). This issue is particularly significant because refusal to participate could introduce an imbalance in socioeconomic demographics; thus, allowing people to opt out of a study could mean losing vital information from populations that researchers have the greatest need to study. Regulations should aim at minimizing these problems instead of exacerbating them.

The advent of genetic biobanks, sample banks, and human microbiome research has created further dilemmas.[5] Many biobanks and sample banks obtain consent from those who donate their samples. When samples are collected, neither researchers nor donors can know the full extent of the research projects that will be performed using the samples. Although donors can give

[5] These issues are discussed in further detail in chapter 6 of this volume.

blanket consent for future research use of their samples, they cannot at that time provide meaningful informed consent to any specific future studies because they have no information about those future studies. When blanket consent is obtained, biobanks and sample banks are unable to conform with several of the general requirements for informed consent enumerated in the Common Rule at 45§46.116a&b. Although the Common Rule already permits IRBs to waive requirements for informed consent when "[t]he research involves no more than minimal risk to the subjects" and when "[t]he research could not practicably be carried out without the waiver or alteration," IRBs can be reluctant to grant a waiver, particularly for studies involving genetic material (Code of Federal Regulations 1991).

To avoid the informed consent problem, some centers resort to anonymizing samples (Clayton 2005; Moros and Rhodes 2010). This makes it nearly impossible to reidentify the sample donors, thereby protecting donor privacy. Although anonymizing allows investigators to use the materials without obtaining informed consent, it also means that investigators are unable to match samples with donors' complete medical records or contact donors again when doing so would enhance research aims. Such measures obliterate some of a sample's associated phenotype information, and thereby severely diminish its scientific value.

In light of these considerations, we suggest that a new category of research risk be created, *de minimis* risk. The U.S. Code of Federal Regulations and some IRB guidance documents refer to three categories of research risk: (1) minimal risk, (2) a minor increase over minimal risk, and (3) more than a minor increase over minimal risk. According to 45 CFR 46.102 (i), "*Minimal risk* means that the probability and magnitude of harm or discomfort anticipated in the research are not greater in and of themselves than those ordinarily encountered in daily life or during the performance of routine physical or psychological examinations or tests." For example, a study involving only a blood draw would count as minimal risk because the risks of the procedure are no greater than the risks associated with a routine physical examination. We use the term *de minimis risk* to indicate an even lesser degree of risk than is currently considered minimal. This new category would apply to studies involving only negligible physical, social, or psychological risk where nothing inherently dangerous is done to the body. Obtaining informed consent should **not** be an absolute requirement for studies that involve only *de minimis* risk.

Our recommendation would make permission to conduct a study without informed consent the default position for studies that involve only a *de minimis* risk; oral or blanket agreement may still be obtained when investigators and/or reviewers determine that obtaining it is feasible and the burdens involved would be reasonable. For example, obtaining a cheek swab or using a leftover blood sample involves only *de minimis* risk. In most circumstances, explaining what is being done and allowing the person to opt out of providing the sample is entirely feasible because the person is present and there are no significant burdens entailed by eliciting agreement. Because the effort involved in explaining and allowing for opting out is reasonable and feasible, obtaining agreement for cheek swab studies, for example, is justified. In a time-sensitive study that involves unconscious trauma patients, the default position of not requiring informed consent or even agreement would apply because the risks involved are *de minimis*.

Our recommendation also incorporates a requirement for balancing risks and benefits. The three most important historical documents that articulate standards of research ethics explicitly endorse a view that research risks should be balanced against the societal benefits that the project promises. For example, Principle 6 of the 1947 Nuremberg Code states that "The degree of risk to be taken should never exceed that determined by the humanitarian importance of the problem to be solved by the experiment" (Nuremberg Code. 1947). Similarly, in the original 1964 version of the Declaration of Helsinki, Basic Principle 4 states that "The importance of the objective is in proportion to the inherent risk to the subject," and Basic Principle 5 states that "Every clinical research project should be preceded by careful assessment of inherent risks in comparison to foreseeable benefits to the subject or to others" (World Medical Association 1964). Also, the Belmont Report discusses the importance of the "Assessment of Risks and Benefits" in Part C, Application 2 (National Commission 1979); and the Common Rule, 45CFR46.111a2, Criteria for IRB approval of research, provides that "Risks to subjects are reasonable in relation to anticipated benefits, if any, to subjects, and the importance of the knowledge that may reasonably be expected to result" (Code of Federal Regulations. 1991).

Nevertheless, the current rules often seem to ignore the importance of adopting a balanced approach. Instead, they focus narrowly on protecting research participants from any risks, regardless of how unlikely, fleeting, or trivial the anticipated harm. When the risks involved are negligible and unlikely, and the study promises to provide a societal benefit, a reasonable assessment should conclude that the balance favors promoting scientific advance. Policies that consider just the risks, and deliberately ignore the possible social benefits that research could provide, express a distorted view of what ethics entails and, therefore, produce regulations that are ethically flawed.

The category of *de minimis* risk would apply to a number of kinds of research and play a role in policy governing the conduct of such studies:

Exempted Research: The Common Rule, 45CFR46.101b, already implicitly employs such a standard when it exempts several kinds of studies from the regulations (Code of Federal Regulations. 1991). This section of the Common Rule exempts educational research, food evaluation, research involving existing data or specimens when they are publicly available, and research involving existing data when they cannot be identified or linked to the subjects. These studies all involve only *de minimis* risk. Our proposal would include all of these already exempt studies under the new *de minimis* risk research category, thereby making the reason for the exemption explicit.

Research on Populations: Public health studies, as well as QI and QA studies that do not involve direct interference with participants' bodies (e.g., sampling effluent from a community), should be considered *de minimis* risk. Because of the vanishingly small likelihood of risk, informed consent should not be required for these activities whenever general participation is needed and when obtaining agreement from individuals is not feasible.

That said, investigators in this age of genetic and human microbiome research will increasingly need to rely on community trust and education to advance their work. Investigators and their institutions need to establish transparent processes for community consultation to review the risks and benefits of research conducted within the community. Such processes are not currently required by regulations governing human subject research, but community consultation is an effective method for establishing trust. It can also play a role in educating communities about the value of research and encouraging the broad participation that is critical to valid and reliable studies.

Biobank and Sample Bank Studies:[6] Even when the process of sample collection itself involves more than *de minimis* risk, subsequent use of samples in biobank and sample bank studies involves only *de minimis* risk of physical harm. The only foreseeable harms that we have identified related to using already collected samples concern the possible social and associated psychological harms from allowing legal proceedings, insurers, family members, employers, or others to violate confidentiality constraints and gain access to biobank materials. NIH certificates of confidentiality have been developed to address that need. Yet, some have argued that the certificates provide inadequate protection (Beskow et al. 2008; Curie 2005; Hermos and Spiro 2009; Melton 1990). An effective mechanism for protecting samples from legal proceedings and other illegitimate access has to be established. When microbiome biobanks and sample banks are safeguarded from use in criminal investigations, immigration proceedings, insurance markets, and the like, and from confidentiality violations that could involve sharing personal information with family members, neighbors, employers, or the like, studies using their samples will involve only *de minimis* risk. Directly responding to concerns about the confidentiality of biobanks with stronger protections is a coherent and effective way of dealing with the problem. It is also a far

[6] Related biobank issues are discussed in this volume's chapter 6. Related confidentiality issues are discussed in chapter 4 of this volume.

better alternative than having investigators either sacrifice the value of samples by anonymizing them or invoking the regulation's definition of "human subject" to redescribe biobank and sample bank studies as something other than human subject research (e.g., human nonsubject research) (Brothers and Clayton 2010).[7]

Furthermore, to directly address the problems associated with obtaining informed consent, regulatory requirements for biobank studies should be adjusted to incorporate an exemption from informed consent. This can most easily be accomplished by classifying further research uses of already obtained samples as *de minimis* risk research. This is true even for studies using de-identified samples linked to medical records (as opposed to anonymized samples that can no longer be linked to a specific human source). Because the physical risks involved in using these samples are only *de minimis*, and because recontacting sample donors to obtain study-specific informed consent is costly, burdensome, sometimes unwelcome, and sometimes unfeasible, informed consent for subsequent studies should not be required.

Discarded Biological Samples: In the course of clinical care, biological samples are routinely collected for analysis. The material that remains after its clinical purpose has been accomplished is often discarded. Some of these otherwise discarded samples will be valuable in various lines of research. The risks involved in using these samples are only *de minimis*, because nothing additional is done to the body of the sample donor. Because it is feasible to allow patients to opt out of the future research use of their samples at the point when samples are collected, and because providing the opportunity to opt out would make the process transparent and more acceptable to patients, institutions should adopt an opt-out policy for the research use of remaining biological samples. Such a practice would allow remaining de-identified samples linked to medical records from patients who did not refuse to participate to be used in research without explicit informed consent.

This proposal for establishing *de minimis* risk as a new category of research risk has several advantages. It puts the focus of research ethics where it belongs, on an assessment of risks and benefits. It avoids the confusion engendered by relying on the stipulative intention-based definition of research that leaves public health agencies and institutions with irresolvable dilemmas. A *de minimis* risk category also reduces unreasonable obstacles that have, until now, inhibited research.

Allowing studies to proceed under the category of *de minimis* risk would strike a reasonable balance between advancing biomedical science and societal health and the importance of respecting persons. Studies that fit this category

---

[7] Greg Koski gave the example of someone who wants to cut down a tree. The fellow hires a plane to seed the clouds, in hopes that lightning will strike and knock over the tree. Koski's point was that this is a very roundabout way to achieve a goal that could be addressed more directly (Koski 2012).

would be exempt from informed consent requirements because the risks involved are truly *de minimis*. Establishing this new category of risk, however, would not be granting blanket exemption from obtaining participants' agreement or blanket consent. The importance of each study, the feasibility of obtaining agreement, and the risks to the participants would still need to be assessed and balanced. When agreement or blanket consent can be obtained with a reasonable effort, investigators should obtain it even when the studies involve only *de minimis* risk.

Additional Factors in the Ethical Conduct of Population Studies

Everyone should do a fair share in contributing to the social benefits provided by public health studies. The benefits of public health studies are widely shared and can be significant, and risks and harms to individuals involved in these studies are typically no more than minimal. Thus, we should treat low-risk studies of the human microbiome as we now treat public heath surveillance and not require informed consent. As we explained earlier, all scientific studies, including public health surveillance, QI, and QA, involve formulating a hypothesis, designing a procedure to test the hypothesis, collecting data, analyzing the data, and drawing conclusions from that analysis. Thus, there is no scientific distinction between research, on the one hand, and surveillance, QI, and QA, on the other.

It makes good sense to treat studies of clinical interventions that involve significant burdens and risks differently than studies in which no or minimal burdens or risks are imposed. Additional factors that should be considered for the ethical conduct of studies involving populations should include whether or not the study:

- Requires broad participation
- Would be significantly distorted by relying on only consenting individuals
- Promises to prevent or allay significant harms
- Promises to provide significant short-term or enduring benefits
- Promises benefits that would be widely shared

Voluntariness and informed consent should not be required for population studies because the risks of public health surveillance, QI, and QA are typically negligible, opting out by any subjects introduces bias, and obtaining informed consent is often not feasible (Coughlin 2006; Fairchild and Bayer 2004; MacQueen and Buehler 2004; Snider and Stroup 1997). For similar reasons, and to efficiently allocate limited oversight resources, IRB review and oversight should be reserved for studies that involve delivering some intervention (e.g., a chemical) to an individual subject because such studies tend to involve greater risks. Also, the benefits that clinical trials promise will typically be relevant to smaller groups than would public health surveillance, QI, and QA. And, because some measure of oversight promotes public trust as well as the trustworthiness of biomedical science, public health agencies and institutions that engage in QI and QA activities should provide their own processes for review and oversight.

Prevention and Education

The Human Microbiome Project (HMP) provides an opportunity to rethink many aspects of the public health message. The most critical message that the public needs to learn is that microbes are not always our enemies, and they do not always have to be destroyed. Indeed, instead of trying to eradicate the bugs, we need to create a healthy internal environment to nurture and maintain a healthy microbiome.

Public health efforts should first focus on collecting and developing evidence about features of a healthy microbiome that are critical to good health and educating the community on how they can be achieved and maintained. Public health agents will then have to determine which information the public needs about microbiome health. They will also have to mount efforts to communicate that information along with the public health message that a healthy microbiome is essential to good health and disease prevention.

Public health agents will also have to devote surveillance efforts to the identification of dangerous elements that can infect the microbiome and track their spread in the population (Schadt and Björkegren 2012). Using rapid genome sequencing for sewage-based epidemiology to produce what Eric Schadt has termed "disease weather maps" could allow public health officials to track dangerous pathogens (Kolata 2011). When a dangerous microbial element is identified, they will also need to identify effective interventions. As public health planning is under way, officials will have to consider which information needs to be disseminated and determine who needs to know what, when, and why. Public health messages should, of course, be directed by scientific evidence, not politics. Messages should transparently communicate accurate information along with clear directions that are supported by a cogent rationale about how to respond.

Communication alone is often not sufficient to achieve effective disease prevention. Prevention includes efforts to reduce the likelihood of adverse health outcomes. It utilizes biomedical, epidemiological, and information technology knowledge to design and implement effective societal conditions and actions aimed at reducing negative consequences (Briss et al. 2000; Graham 2002; Lavis et al. 2004, Moynihan et al. 2006; World Health Organization 2003).

Knowledge emerging from the study of the microbiome provides an opportunity to draw on what has been learned about prevention. To improve health outcomes for individuals and populations, effective prevention programs related to smoking, human immunodeficiency virus (HIV)/acquired immunodeficiency syndrome (AIDS), cardiovascular disease, and other conditions have been implemented. Multidisciplinary groups of investigators have elucidated critical questions about individual and collective health-seeking behavior, cultural competency, and health literacy, all of which are essential for designing effective public health messages and motivating people to act on the information. Experts in prevention have developed an understanding of behavioral and environmental factors that put individuals and populations at different levels of risk. That information enables public health educators to tailor their messages and develop links with the affected communities to engage at-risk populations in risk reduction (Shiffman et al. 2005). Integrating what has been learned about averting or reducing negative disease consequences related to the microbiome into a prevention agenda will facilitate the work of public health agents. It will enable them to apply established models to the communication of actionable behavior. Public health agents can thereby reduce the negative consequences of microbial threats by focusing on encouraging health-promoting behaviors, reducing unhealthy and dysfunctional behaviors, and promoting psychological resilience.

Ecology has taught us about the interconnectedness of the environment and health. Epidemiology has provided us with a methodology for analyzing and improving our understanding of populations. Experts from related fields have contributed their insights into the impact of group behavior on individual health outcomes. Linking the growing body of knowledge about the microbiome with what we already know about prevention will facilitate the implementation of effective public health interventions.

Cautions for Public Health Policymakers

As we learn more about the microbiome, we will also want to develop public health surveillance programs, disease prevention policies, and educational campaigns aimed at promoting healthy microbiomes and preventing diseases associated with microbes. As a general caution, medical and public health policymakers need to be alert to the kinds of illegitimate considerations that can distort and pervert any policy. In that light, consider some of the public health policies enacted in our recent past.

Common psychological tendencies can interfere with judgment. Human psychology inclines people to exaggerate the impact of a loss and underappreciate the value of future goods (Kassam et al. 2008). Consider some public health interventions from the fall of 2001. In the face of a conceivable but remote and unlikely smallpox threat, exaggerated risk aversion inclinations may have contributed to the hasty and risky vaccination policies after 9/11. Similarly, politics and self-interest can influence public policy. It is possible that political goals, cronyism, and fear all contributed to the huge investment in anthrax decontamination in the fall of 2001 at sites where the likelihood of danger was almost nonexistent. Those biasing elements may also have contributed to the considerable investments in federal funding for decontamination equipment in the years that followed.

Similarly, the desire to curry favor or to shore up votes does not promote reasonable public health policy, and neither does greed, which can be camouflaged under seemingly acceptable justifications. Greed coupled with the desire to repay debts or to secure future political support may have contributed to the decision to administer unproven anthrax vaccination to U.S. troops after 9/11. Prejudice, stereotyping, the desire to do something, pressing needs made vivid by individual cases, lack of insight, and lack of foresight are other common psychological inclinations that can distort judgment and lead to unjust public health policies (Rhodes 2005).

Our experience post-9/11, the flu vaccine shortage, and dealing with Hurricane Katrina have also taught us lessons about communication and trust and their importance in the design and implementation of just public health policies. After the fall of the World Trade Center Twin Towers, public health officials from the Environmental Protection Agency and government representatives failed to honestly communicate about the environmental dangers, and they misinformed the public about the air quality and the need for protection from the toxic environment. Today, thousands of people who worked at the site are ill and hundreds are dying, at least in part because of the failures to provide full and honest disclosure (DePalma 2006, 2007a, 2007b). Apparently, those who made the decisions to withhold information and to promulgate false reports were more concerned with impressions, messages, and promoting political ends than with promoting safety. In a similar way, false and misleading reports before, during, and after Hurricane Katrina cost lives and exaggerated the tragedy. These inaccurate and misleading communications undermined trust in government, in public health pronouncements, and in public health policy (Cooper and Block 2006). When people believe that they are being deceived and that policies favor personal or political advantage rather than the public good, they are less inclined to accept the pronouncements and less likely to cooperate with the policy.

In contrast, the honest communication about the 2004 flu vaccine shortage and the clear communication about justification for the policies that governed the distribution of the scarce supply contributed to cooperation with the policy and the resulting success in avoiding deaths and serious illness (DesRoches, Blendon, and Benson 2005). These examples highlight the need for full and honest communication about medicine and matters of public health and the importance of communication in promoting justice.

Creating good public policy is always a risky business because changes lead to many consequences. It is hard to foresee all of the consequences of any change and impossible to calculate all of the effects any policy will initiate. Changing existing policy and implementing new policy unavoidably involves uncertainty. This is especially true when policy is crafted in response to early research in a new field of study such as the human microbiome. Although we are beginning to appreciate the ways in which the human microbiome may contribute to disease resistance and susceptibility and have a significant impact on public health, caution is in order for those crafting public health surveillance or interventions. Policymakers will have to maintain their focus on the appreciation of health and safety, on the one hand, and liberty, on the other. Both are important social values.

Caution will be needed to avoid overselling the significance of findings in microbiome research. It will also be critical to avoid the temptation to do something before the risks of an intervention are adequately assessed. Before implementing any measures to protect public health, it will be important to accurately assess and compare the harms and benefits of proposed interventions. The difficulty of comparing alternative courses is complicated by the problem of comparing risks of different kinds and magnitudes, as well as the ways in which these risks will affect different individuals. Given the inherent complexity of what is being assessed, the incommensurability of the consequences that must be compared, and the obstacles in determining the sum total of the positive and negative results of any policy, rationality and care should be the guiding forces in forging public health policy. We are not going so far as to embrace the precautionary principle that opposes any intervention until it can be proven not to be harmful. Rather, we urge caution, reason, and vigilant attention to the invisible influences that can distort good judgment.

At least two issues of public health microbiome policy have already become contentious. Obesity has been recognized as a serious public health problem that increases the risk of diseases such as diabetes and heart disease and decreases life expectancy. The proportion of the U.S. population with obesity appears to be growing, especially among the poor and African Americans (Wyatt, Winters, and Dubbert 2006). Data from early studies seem to implicate the human microbiome in obesity, although causality has not yet been established (Ley 2010). Public health surveillance may be necessary to track the phenomenon. At some point, intervention may be indicated to combat the condition, and then important questions will have to be answered. How strong must the evidence be before intervention is justified? Can a dramatic infringement on liberty be justified to improve the health of people who turn out to have an unhealthy microbiome? Will declaring that the obese have an unhealthy microbiome stigmatize them and encourage discrimination? Will the justification for an intervention turn on how communicable unhealthy microbiomes turn out to be?

A second issue that already has generated public health policy in Europe and the United States involves providing antibiotics to cows and other farm animals. The European Union (E.U.) has banned the prophylactic use of antibiotics for milk cows out of concern for the spread of antibiotic resistance to humans through their milk (Selgelid 2007). We can ask whether the available evidence of this harm is strong enough to justify that policy. We can also consider whether the health risks of preventing antibiotic resistance are significant enough to outweigh the possible economic consequences to farmers. We should also take into account whether cautions about milk will make parents fearful of offering milk to their children and thus negatively affect nutrition. Furthermore, we should consider whether withholding antibiotics will raise the cost of milk enough to make this source of nutrition unavailable to the poor and, thereby, make some of the least advantaged people worse off. Most critically, however, we will have to ask with respect to these and any other concerns that may be raised whether there is adequate supporting evidence that justifies forging policy (Atkins et al. 2004; Fretheim, Schünemann, and Oxman 2006; Oxman, Fretheim, and Schünemann 2006).

Although it is hard to compare the risks and benefits in light of the inherent complexities and uncertainties, the evidence to support restriction on prophylactic antibiotics is mounting. For example, Lance Price and colleagues report on their study in the February 2012 issue of *mBio*, the journal of the American Society for Microbiology. They show how *Staphylococcus aureus* CC398, one of the bacterial strains associated with methicillin-resistant *S. aureus* (MRSA) infections, "originated in humans as MSSA [methicillin-susceptible *S. aureus*] and then spread to livestock, where it acquired resistance to methicillin and tetracycline" (Price et al. 2012). The authors argue from their genomic analyses and previous epidemiological data that "livestock-associated CC398 has been linked to an increase in MRSA infections in northern Europe." The impact of these developments will require additional research and surveillance to determine their public health impact.

On January 4, 2012, the U.S. Food and Drug Administration (FDA) announced new restrictions on antibiotic use in farm animals. Like many issues in medicine, this one is not completely settled. People who are supported by agricultural or pharmaceutical interests argue that fears are exaggerated and there is insufficient evidence to support restrictive policy (*New York Times* 2012). At the same time, others see the need for more restrictions. For example, Representative Louise M. Slaughter, a Democrat from New York and a microbiologist, is quoted in the *New York Times* complaining that the FDA's response had not been timely or aggressive enough. According to Slaughter, "We are staring at a massive public health threat in the rise of antibiotic-resistant superbugs.... We need to start acting with the swiftness and decisiveness this problem deserves" (Harris 2012). While some maintain that E.U. policy may be overly restrictive, documented cases of resistant infections acquired from animals date back over 30 years. There are now concerns about antibiotics in manure that is spread on farmland and gets into water supplies (Lyons 1982; Lyons et al. 1980). Furthermore, preliminary evidence from Denmark suggests that the E.U. restrictions have not significantly affected milk production (Wegener 2003).

Taking a broad view of the consequences of any public health intervention requires calculation of all of the costs of the intervention and all of the costs of the outcomes related to the decision. These costs include the financial costs of unfunded mandates as well as the opportunity costs of other possible uses for the funds that would be diverted into the implementation of the policy. Crafting public health interventions for the purpose of maintaining healthy microbiomes can be useful and beneficial, but it will require cautious maneuvering as well as moral imagination and creativity.

To add a practical perspective on the importance of public health and population research, we include the following commentary.

Coda: Further Philosophical Reflections on Public Health and the Microbiome
by William James Earle

Methodological individualism tells us that only individuals can be healthy or sick. Only individuals can have beliefs, desires, fears, and preferences. Only individuals can hope. Only individuals can die. Not much follows from this platitude, but something does.

On the individual level, it is the patient who goes to the doctor, not the doctor who goes to the patient. Does this mean it is the patient who decides when he or she is not well? Yes and no. The situation is complicated, even if the complications can generally be ignored in the practice of medicine where concepts of *sickness*, *health*, *disease*, and *pathology* are likely to be (and can be) taken for granted. More precisely, the italicized items in the previous sentence are words that, though they can be part and parcel of ordinary fluency, may not be connected to entirely transparent or trouble-free concepts.

The basic problem is simply that the laws of nature, unlike the laws of the land, cannot be broken and that the laws of nature apply to organisms, to us, for example, as they do to everything else. Put another way, nature is indifferent as between HIV and humans, or between erythrocytes and leukemia cells. A child who dies a painful death dies, not because the laws of nature are broken, but because they are inexorable, inevasible. There are no "miracle deaths" for the same reason that there are no "miracle cures" (Earle 2010).

This is the theme of Georges Canguilhem's *Le normal et le pathologique*. To paraphrase him, pathology arises when, and only when, alteration of the normal state is experienced as undesirable, as negative. In the passage where Canguilhem discusses this concept he references a conscious individual, but he thinks that something like what we feel (and can reflect on and articulate) must be projected across the whole range of organisms. He writes, "To live, even in the case of an ameba, is to prefer and exclude." An interest in survival is a defining feature of an organism and organisms are hard-wired to survive, even in the case of humans who have the power perhaps unique in the animal kingdom to short-circuit the wiring.

The pathological, Canguilhem writes, implies pathos, the immediate and concrete sentiment of suffering and powerlessness, of life thwarted (Canguilhem 1972). Of course, as medicine is actually practiced, it must address itself, with benign presumption, to all the silent or secret precursors of conditions that will be disliked and regretted when they emerge. It has also to be assumed that people desire longevity, and that no one, save perhaps suicide bombers, pays much attention to Socrates' remark that it is more important how one lives than how long one lives.

Here's why I think the Canguilhem considerations are relevant. They allow us think about autonomy and paternalism in a manner interestingly orthogonal to an approach based on rights and respect. We can say: we should really respect the views of the patient, but we can say something stronger: those views provide the only conceptually available basis for identifying pathology. This is metaphysics and will most of the time leave presumptive practice alone. There are nevertheless occasions when awareness of the metaphysical background will help us to achieve clarity.

Public health issues are embedded in fundamental questions of political economy. This is much-canvassed territory and I offer only a few stray thoughts. We must ask what a government should do for its citizens and other residents within its national territories. Sometimes this is answered by saying a government should only do what the free market cannot do. This gives us things like national defense, the FBI, the CIA, and many others things that, as a matter of fact, are paid for by taxes, federal, state, and local.

This is, of course, a lousy answer. All schools could be private. All schools could be public. Much of the security provided by police could be privatized and much of it has been privatized. Estimates vary but agree that, in the United States, contract security officers already outnumber sworn law enforcement officers.

The good question is, not what is possible, but what is desirable. This shifts us to a fairly philosophical register: What counts as a decent society? What kind of society do we want to live in? What kind of lives do we want to lead? Societies differ, whether people seem friendly or hostile, whether they are gregarious or keep to themselves, whether they are gentle or violent, trusting or suspicious.

Could all research, all research and development (R&D), be private? Could all biomedical research be private? Yes to both questions. But then our economy would be worse, our health would be worse, and we would move in the direction of intellectual mediocrity. Part of the issue is long lead time. Private research tends, certainly not always but often enough, to concentrate where there is a reasonable hope of quick commercial exploitation. The Human Microbiome Project, undertaken with federal funding through the NIH, quite apart from its intrinsic interest (that all men by nature desire to know), will certainly yield actionable knowledge, but not at once.

These thoughts are very indirectly related to the Human Microbiome Project. But there is at least one thing that can be picked up on. Tony Robert Judt says that Americans do not like, or at least think they do not like, an interventionary state (Judt 2009). A chubby friend of mine recently remarked that he did not like the idea of fast food outlets being required to post calorie counts. This was, he said, an example of the nanny state. I am an adult, I guessed him to be thinking, and I don't want the state telling me what to eat. But the state was not telling him anything. At most it was suggesting or attempting, in a relatively quiet way, to be persuasive. The provision of information, as such, is not coercive, even in the unusual case where it is bound to affect action. The informational placard in the fast food outlet is an instance of what Richard Thaler and Cass Sunstein refer to as libertarian paternalism. They explain this apparent, but merely apparent, oxymoron as follows: "The libertarian aspect of our strategies lies in the straightforward insistence that, in general, people should be free to do what they like and to opt out of undesirable arrangements if they want to do so" (Thaler and Sunstein 2008, 5). The paternalistic aspect lies in the claim that it is legitimate for choice architects to try to influence people's behavior to make their lives longer, healthier, and better. Neither half of Thaler and Sunstein's apparent oxymoron is altogether trouble free. When we recommend that people should be allowed to do what they like, or what (to use the economist's standard term) they prefer, we have to consider that people often do not know what they prefer. The economic theory of revealed preference sidesteps this problem, but it is serious and real. Preference formation itself, like belief formation, can be pathological. For example, an entirely real and authoritative preference can be formed through social mimicry. Antipaternalists, Thaler and Sunstein remark, typically assume, incorrectly, that almost all people, almost all of the time, make choices that are in their best interest or at the very least are better than the choices that would be made by someone else. No one needs to be an expert on the full spectrum of the consequences, especially remote consequences of his or her actions. Two perceptive (and characteristically elegant) remarks of John Dunn are worth quoting:

> Do human beings really know what they want? As to what I want myself, I am no doubt the world's leading authority; but the incessant pressures of regret certainly establish that even I am an eminently fallible authority.

> The immediacy and authority of present desire is more than offset, for purposes of political alignment and action, by its rapid evanescence. The view that every person is best placed to assess their own interests is therefore better seen as a somewhat bleary prudential reminder of the hazards of permanently alienating such assessment to any other person or institution than as a defensible axiom in a theory of the human good. (Dunn 1984, 32)

The idea of a choice architect makes me nervous. Still, there seem to be many instances of choice architecture where nothing nefarious is going on. I remember being puzzled some years ago in the men's room at Schiphol Airport in Amsterdam by the image of a black housefly etched on each urinal. It struck me as an odd decorative touch. Thaler and Sunstein provide the explanation: It seems that men usually do not pay much attention to where they aim, which can create a bit of a mess, but if they see a target, attention and therefore accuracy are much increased. An economist named Aad Kieboom, who happened to be in charge of the airport's expansion, came up with the idea and reports that the etchings reduced spillage by 80%. This is an example of a nudge, and I cannot imagine grounds for objecting to it. No one is being manipulated or coerced away from an initial preference for sloppiness (Thaler and Sunstein 2008, 4).

Here's another benign example and one that is closer to issues in public health. Would anyone object to putting the fruit and salad before the desserts in an elementary school cafeteria if the result were to induce kids to eat more apples and fewer Twinkies? It should be noted that nudges operate behind the backs of agents like the hidden persuaders that Vance Packard worried about in his eponymous 1957 book (Packard [1957] 2007). In contrast, an informational placard, like the warning on a cigarette pack, addresses the rational agent. This is generally supposed to be a good thing, but information provided by public health authorities has (1) to be noticed, (2) to be understood and remembered, and (3) to be responded to rationally. There is plenty of room for each of these things not to happen.

A benign choice architect seeks to make our lives longer, healthier, and better. Better is especially worrisome. Lives would be better, in my judgment, with much less TV watching, and indeed many people say they wished they watched less TV. Nevertheless, I shouldn't want to attempt in any public way to regulate or curtail TV watching. Of course, as a parent and choice architect, one can try to limit TV watching by filling one's children's lives with more appealing options.

Imagine if a 12-step program or some other effective behavior-modifying regimen had gotten hold of Dylan Thomas, Hart Crane, and Malcolm Lowry. Their lives might have been longer and healthier but, with their genius normalized out of them, not necessarily and probably not better.

Policy Recommendations

- Because there is no scientific distinction between research, on the one hand, and observational surveillance, quality assurance, and quality improvement, on the other, these names should not be the basis for regulating them differently.
- Relevant factors that should be considered in establishing standards and rules for the ethical conduct of studies that involve populations of human subjects include whether or not the study:
  - Imposes significant or likely enduring harms (or none)
  - Requires broad participation
  - Would be significantly distorted by relying on only consenting individuals
  - Promises to prevent or allay significant harms
  - Promises to provide significant short-term or enduring benefits
  - Promises benefits that would be widely shared
- Samples collected for biobanks, clinical purposes, surveillance, quality assurance, or quality improvement can be very useful in studying or preventing disease. Because the risks to individuals from using these samples is very low, we should consider all such studies to involve only *de minimis* risk and not require informed consent when obtaining it would be unfeasible or unreasonably burdensome.
- Public health agencies and institutions that engage in *de minimis* risk research, including surveillance, quality assurance, and quality improvement activities, should employ a process for review and oversight.
- Public health policies should be supported by carefully evaluated evidence. The justification for policies should be honestly communicated to the public so as to maintain the communities' trust and assure cooperation with implementation.

References

Atkins, D., D. Best, P. A. Briss, M. Eccles, Y. Falck-Ytter, S. Flottorp, G. H. Guyatt, R. T. Harbour, M. C. Haugh, D. Henry, S. Hill, R. Jaeschke, G. Leng, A. Liberati, N. Magrini, J. Mason, P. Middleton, J. Mrukowicz, D. O'Connell, A. D. Oxman, B. Phillips, H. J. Schunemann, T. T. Edejer, H. Varonen, G. E. Vist, J. W. Williams, Jr., and S. Zaza. 2004. Grading quality of evidence and strength of recommendations. *BMJ* 328(7454): 1490.

Baily, M. A. 2008. Harming through protection. *New England Journal of Medicine* 358(8): 768–769.

Baily, M. A., M. M. Bottrell, J. B. Lynn. 2006. The ethics of using QI methods to improve health care quality and safety. *Hastings Center Report* 36: S1–S40.

Battin, M. P., L. P. Francis, J. A. Jacobson, and C. B. Smith. 2009. *The Patient as Victim and Vector: Ethics and Infectious Disease.* New York: Oxford University Press.

Beskow, L. M., L. Dame, and E. J. Costello. 2008. Certificates of confidentiality and the compelled disclosure of research data. *Science* 322(5904): 1054–1055.

Betancourt, J. R. and R. K. King. 2003. Unequal treatment: the Institute of Medicine report and its public health implications. Public Health Rep. 2003 Jul–Aug; 118(4): 287–292.

Boyle, Jr., J. M. 1980. Towards understanding the principle of double effect. *Ethics* 90(4): 527–538.

Briss, P. A., S. Zaza, M. Pappaioanou, J. Fielding, L. Wright-De Aguero, B. I. Truman, D. P. Hopkins, P. D. Mullen, R. S. Thompson, S. H. Woolf, V. G. Carande-Kulis, L. Anderson, A. R. Hinman, D. V. McQueen, S. M. Teutsch, and J. R. Harris. 2000. Developing an evidence-based Guide to Community Preventive Services—methods. The Task Force on Community Preventive Services. *American Journal of Preventive Medicine* 18(1 Suppl): 35–43.

Brothers, K. B., and E. W. Clayton. 2010. "Human non-subjects research": Privacy and compliance. *American Journal of Bioethics* 10(9): 15–17.

Canguilhem, G. 1972. *Le normal et le pathologique*. Paris: Presses universitaires de France.

Centers for Disease Control and Prevention. 1999. *Guidelines for Defining Public Health Research and Public Health Non-Research*. Revised October 4. http://www.cdc.gov/od/science/integrity/docs/defining-public-health-research-non-research-1999.pdf. Accessed April 13, 2013.

Code of Federal Regulations. 1991. Federal Policy for the Protection of Human Subjects The Federal Policy for the Protection of Human Subjects, (the "Common Rule") Title 45, Part 46. Protection of Human Subjects. Rockville, MD: U.S. Department of Health and Human Services. http://www.wma.net/en/30publications/10policies/b3/17c.pdfhttp://www.hhs.gov/ohrp/humansubjects/guidance/45cfr46.html Accessed April 3, 2013.

Cheney, R. A. 1984. Seasonal aspects of infant and childhood mortality: Philadelphia, 1865-1920. *Journal of Interdisciplinary History* 14: 561–585.

Clayton, E. W. 2005. Informed consent and biobanks. *Journal of Law, Medicine and Ethics* 33(1):1073–1105.

Cohen, A. B., J. D. Restuccia, M. Shwartz, J. Drake, R. Kang, P. Kralovec, S. K. Holmes, and D. Bohr. 2008. A survey of hospital quality improvement activities. *Medical Care Research and Review* 105: 1–2.

Cooper, C., and R. Block. 2006. *Disaster: Hurricane Katrina and the Failure of Homeland Security*. New York: Henry Holt and Company.

Coughlin, S. 2006. Ethical issues in epidemiologic research and public health practice. *Emerging Themes in Epidemiology* 3(16): 1–10.

Currie, P. M. 2005. Balancing privacy protections with efficient research: Institutional review boards and the use of certificates of confidentiality. *IRB: Ethics and Human Research* 27(5): 7–12.

DePalma, A. 2006. Illness persisting in 9/11 workers, big study finds. *New York Times*, September 6.

DePalma, A. 2007a. As a way to pay victims of 9/11, insurance fund is problematic. *New York Times*, February 15.

DePalma, A. 2007b. Ground zero illnesses clouding Giuliani's legacy. *New York Times*, May 14.

DesRoches, C. M., R. J. Blendon, and J. M. Benson. 2005. Americans' responses to the 2004 influenza vaccine shortage. *Health Affairs* 24(3): 822–831.

Dunn, John. 1984. *The Politics of Socialism: An Essay in Political Theory*. Cambridge: Cambridge University Press.

Dupuis, E. M. 2002. *Nature's Perfect Food: How Milk Became America's Drink.* New York: New York University Press.

Earle, W. J. 2010. Public health. Proceedings of the NIH Panel at Mt. Sinai Hospital. http://www.williamjamesearle.net/. Accessed January 21, 2010.

Emanuel, E. J., and J. Menikoff. 2011. Reforming the regulations governing research with human subjects. *New England Journal of Medicine* 365: 1–6.

Emanuel, E. J., A. Wood, A. Fleischman, et al. 2004. Oversight of human participants research: identifying problems to evaluate reform proposals. *Annals of Internal Medicine* 141: 282–291.

Fairchild, A. L., and R. Bayer. 2004. Ethics and the conduct of public health surveillance. *Science* 3: 631–632.

Feinstein, A. R., and R. I. Horwitz. 1978. A critique of the statistical evidence associating estrogens with endometrial cancer. *Cancer Research* November 38(11 Pt 2): 4001–4005.

Ferreira-Gonzalez I., J. R. Marsal, F. Mitjavila, A. Parada, A. Ribera, P. Cascant, N. Soriano, P. L. Sanchez, F. Arós, M. Heras, H. Bueno, J. Marrugat, J. Cuñat, E. Civeira, and G. Permanyer-Miralda. 2009. Patient registries of acute coronary syndrome: assessing or biasing the clinical real world data? *Circulation: Cardiovascular Quality and Outcome* 2: 540–547.

Fink, A. M. 1993. *Classical and New Inequalities in Research.* Boston: Kluwer Academic.

Flanagan, B. M., S. Philpott, and M. A. Strosberg. 2011. Protecting participants of clinical trials conducted in the ICU. *Intensive Care Medicine* 26(4): 237–249.

Foot, P. 1967. The problem of abortion and the doctrine of double effect. *Oxford Review* 5: 5–15.

Fost, N., and R. J. Levine. 2007. The dysregulation of human subjects research. *Journal of the American Medical Association* 298(18): 2196–2198.

Fretheim, A., H. J. Schünemann, and A. D. Oxman. 2006. Improving the use of research evidence in guideline development: 3. Group composition. *Health Research Policy and Systems* 4: 15.

Gawande, A. 2007. A lifesaving checklist. *New York Times,* December 30.

Graham, I. D., M. B. Harrison, M. Brouwers, B. L. Davies, and S. Dunn. 2002. Facilitating the use of evidence in practice: evaluating and adapting clinical practice guidelines for local use by health care organizations. *Journal of Obstetrics, Gynecology, and Neonatal Nursing* 31(5): 599–611.

Harris, G. 2012. Citing drug resistance, U.S. restricts more antibiotics for livestock. *New York Times,* January 4.

Heckman, J. 1979. Sample selection bias as a specification error. *Econometrica* 47: 153–161.

Hermos, J. A., and A. Spiro. 2009. Certificates should be retired. *Science* 323(5919): 1288–1299.

Hodge, J. G., and L. O. Gostin. 2004. *Public Health Practice vs. Research: A Report for Public Health Practitioners.* Atlanta, GA: CSTE Advisory Committee.

Infectious Diseases Society of America. 2009. Grinding to a halt: the effects of the increasing regulatory burden on research and quality improvement efforts, *Clinical Infectious Diseases* 49(3): 328--335.

Inland Empire Perchlorate Ground Water Plume Assessment Act of 2010, H.R.4252, 111th Cong., 2nd Sess. (2010).

Judt, Tony. 2009. What is living and what is dead in social democracy? *New York Review of Books* LVI(20): 86.

Kass, N., et al. 2003. Controversy and quality improvement: lingering questions about ethics, oversight, and patient safety research. *The Joint Commission Journal on Quality and Patient Safety* 34(6): 349–353.

Kassam, K. S., D. T. Gilbert, A. Boston, and T. D. Wilson. 2008. Future anhedonia and time discounting. *Journal of Experimental Social Psychology* 44(6): 1533–1537.

Kofke, W. A., and M. A. Rie. 2003. Research ethics and law of healthcare system quality improvement: the conflict of cost containment and quality. *Critical Care Medicine* 31(3): S143–S152.

Kolata, G. 2011.The new generation of microbe hunters. *New York Times*, August 29, Science.

Koske, Greg. 2012. The future of human subjects research regulation. Petrie-Flom Center Conference, Harvard Law School, May 18–19.

Kulynych, Jennifer, and David Korn. 2002. The new federal medical-privacy rule. *New England Journal of Medicine* 347(15): 1133–1134.

Langmuir, A. D. 1980. The epidemic intelligence service of the Center for Disease Control. *Public Health Reports* 95: 470–477.

Lavis, J. N., F. B. Posada, A. Haines, and E. Osei. 2004. Use of research to inform public policymaking. *Lancet* 364(9445): 1615–1621.

Ley, R. E. 2010. Obesity and the human microbiome. *Current Opinion in Gastroenterology* 26(1): 5–11.

Lynn, J., M. A. Baily, M. Bottrell, et al. 2007. The ethics of using quality improvement methods in health care. *Annals of Internal Medicine* 146(9): 666–673.

Lyons, R. W. 1982. Animal-to-human disease transmission. *Review of Infectious Diseases* 4(3): 733. PubMed PMID: 7123044.

Lyons, R. W., C. L. Samples, H. N. DeSilva, K. A. Ross, E. M. Julian, and P. J. Checko. 1980. An epidemic of resistant Salmonella in a nursery. Animal-to-human spread. *Journal of the American Medical Association* 243(6): 546–547. PubMed PMID: 7351786.

MacQueen, K. M., and J. W. Buehler. 2004. Ethics, practice and research in public health. *American Journal of Public Health* 94(6): 928–932.

McIntyre, A. 2009. Doctrine of double effect. In Edward N. Zalta (ed.), *The Stanford Encyclopedia of Philosophy* (Fall 2009 ed.). http://plato.stanford.edu/archives/fall2009/entries/double-effect/. Accessed April 13, 2013.

Melton, G. B. 1990. Certificates of confidentiality under the Public Health Service Act: Strong protection but not enough. *Violence and Victims* 5(1): 67–71.

Miller, F. G., and E. J. Emanuel. 2008. Quality-improvement research and informed consent. *New England Journal of Medicine* 358: 765–767.

Moros, D. A., and R. Rhodes. 2010. Privacy overkill, *American Journal of Bioethics* 10: 12–15.

Moynihan, R., A. D. Oxman, J. N. Lavis, and E. Paulsen. 2006. *Evidence-Informed Health Policy: Using Research to Make Health Systems Healthier. A Review of Organizations That Support the Use of Research Evidence in Developing Guidelines, Technology Assessments, and Health Policy, for the WHO Advisory Committee on Health Research.* Oslo: Norwegian Knowledge Centre for the Health Services.

National Commission for the Protection of Human Subjects of Biomedical and Behavioral Research. 1979. *The Belmont Report: Ethical Principles and Guidelines for the Protection of Human Subject of Research.* Washington, DC: Office for Protection from Research.

New York Times. 2012. F.D.A. creeps forward. Editorial. January 10.

Nuremberg Code. II Trials of War Criminals Before the Nuremberg Military Tribunals Under Control Law No. 10, 181-83. U.S. Government Printing Office. 1946–1949. http://www.hhs.gov/ohrp/archive/nurcode.html Accessed, April 3, 2013.

Office for Human Research Protections (OHRP). 2008 *OHRP Statement Regarding the New York Times Op-Ed Entitled A Lifesaving Checklist.* January 15.

O'Herrin, J. K., N. Fost, and K. A. Kudsk. 2004. Health insurance portability accountability act (HIPAA) regulations effect on medical record research. *Annals of Surgery* 239(6): 666–673.

Oliver, J. W. 1956. *History of American Technology.* New York: Ronald Press.

Osler, W. 1899. *The Principles and Practice of Medicine.* New York: D. Appleton and Company.

Oxman, A. D., A. Fretheim, and H. J. Schünemann. 2006. Improving the use of research evidence in guideline development. *Health Research Policy and Systems* 4: 14.

Packard, V. O. (1957) 2007. *The Hidden Persuaders: An Introduction to the Techniques of Mass-Persuasion Through the Unconscious.* Brooklyn, NY: Ig Publishing.

Porter, D. 1999. *Health, Civilization and the State: A History of Public Health from Ancient to Modern Times.* London: Routledge Press.

Price, Lance B., Marc Stegger, Henrik Hasman, et al. 2012. *Staphylococcus aureus* CC398: host adaptation and emergence of methicillin resistance in livestock. *mBio.* doi:10.1128/mBio.00305-11.

Pritchard, I. A. 2008. OHRP Correspondence: Letter to Dr. Peter Provonost Regarding Indwelling Catheter Procedures (pp. 1–4).Washington, DC: U.S. Department of Health and Human Services, Office for Human Research Protection. *http://www.hhs.gov/ohrp/policy/Correspondence/pronovost20080730letter.pdf Accessed April 13, 2013. Pronovost, P., D. Needham, S. Berenholtz, et al. 2006. An intervention to decrease catheter-related bloodstream infections in the ICU. *New England Journal of Medicine* 355: 2725–2732.

Resnik, D. B., R. Sharp, and D. Zeldin. 2005. Research on environmental health interventions: ethical problems and solutions. *Accountability in Research* 12: 69–101.

Rhodes, R. 2005. Justice in medicine and public health. *Cambridge Quarterly of Healthcare Ethics* 14: 13–26.

Rose, G. 1981. Strategy of prevention: lessons from cardiovascular disease. *Medical Practice: British Medical Journal* 282: 1847–1851.

Rose, G. 1992. Epidemiology and environmental risks. *Soz Priventivmed* 37: 41–44.

Ross, W. D. 1930/2002. *The Right and the Good.* New York: Oxford University Press.

Rossi, P. H., and H. E. Freeman. 1993. *Evaluation: A Systematic Approach* (5th ed.). Newbury Park, CA: Sage.

Scanlon, T. M. 2008. *Moral Dimensions: Permissibility, Meaning, Blame.* Cambridge, MA: Belknap Press of Harvard University Press.

Schadt, E. E., and J. L. M. Björkegren. 2012. Network-enabled wisdom in biology, medicine, and health care. *Science Translational Medicine* 4 January 4(115): 115rv1.

Schwab, A. 2010. The recipe for overreaching regulation. *American Journal of Bioethics.* 10(8): 55–56.

Selgelid, M. J. 2007. Ethics and drug resistance. *Bioethics* 21(4): 218–229.

Seto, B. 2001. History of medical ethics and perspectives on disparities in minority recruitment and involvement in health research. *American Journal of the Medical Sciences* 322: 246–250.

Shen, J. J., L. F. Samson, E. L. Washington, P. Johnson, C. Edwards, and A. Malone. 2006. Barriers of HIPAA Regulation to Implementation of Health Services Research. *Journal of Medical Systems* February 30(1): 65–69.

Shiffman, R. N., J. Dixon, C. Brandt, A. Essaihi, A. Hsiao, G. Michel, and R. O'Connell. 2005. The GuideLine Implementability Appraisal (GLIA): development of an instrument to identify obstacles to guideline implementation. *BMC Medical Informatics and Decision Making* 5: 23.

Snider, Jr., D. E., and D. F. Stroup. 1997. Viewpoint on human subjects research. *Public Health Reports* 112: 29–32.

Thacker, S. B., and R. L. Berkelman. 1988. Public health surveillance in the United States. *Epidemiologic Review* 10: 164–190.

Thaler, R. H., and C. R. Sunstein. 2008. *Nudge: Improving Decisions About Health, Wealth and Happiness.* New Haven, CT: Yale University Press.

Torrey, F. E., and R. H. Yolken. 2005. *Beasts of the Earth: Animals, Humans, and Disease.* Piscataway, NJ: Rutgers University Press.

Turnock, B. J. 2007. *Essentials of Public Health.* Sudbury, MA: Jones and Bartlett Press.

Vates, J. R., J. L. R. Hetric, K. L. Lavin, G. K. Sharma, R. L. Wagner, and J. T. Johnson. 2005. Protecting medical record information: start your research registries today. *Laryngoscope* 115: 441–444.

Wegener, Henrik C. 2003. Antibiotics in animal feed and their role in resistance development. *Current Opinion in Microbiology* 6: 439–445.

Wenzel, R. P., and M. B. Edmond. 2006. Team-based prevention of catheter-related infections. *New England Journal of Medicine* 355: 2781–2783.

Winslow, C. E. A. 1920. The untilled fields of public health. *Science* 51: 23–33.

World Health Organization. 2003. *World Health Organization Process for a Global Strategy on Diet, Physical Activity and Health. WHO/NMH/EXR.02.2 Rev.1.* Geneva: World Health Organization.

World Medical Association Declaration of Helsinki. 1964. http://www.wma.net/en/30publications/10policies/b3/17c.pdf Accessed, April 3, 2013.

Wyatt, S. B., K. P. Winters, and P. Dubbert. 2006. Overweight and obesity: prevalence, consequences and causes of a growing public health problem. *American Journal of the Medical Sciences* 33(4): 166–174.

GLOSSARY

16S ribosomal RNA An essential component of the bacterial ribosome that is highly conserved among prokaryotes. Metagenomic studies often use samples of the 16S rRNA gene. Because it is universal and highly conserved, it is an ideal choice for characterizing the diversity among microbes.

Algae Simple microorganisms that contain chlorophyll and can therefore carry out photosynthesis. They live in aquatic habitats and moist environments on land. The algal body may be unicellular or multicellular, filamentous, ribbon-like, or plate-like.

Archaea A domain of prokaryotic organisms that includes archaea bacteria such as the methanogens, which produce methane. Thermoacidophilic bacteria live in extremely hot and acidic environments such as hot springs (see *thermophilic*). Halophilic bacteria can only function at high salt concentrations and are abundant in the world's oceans.

Bacteria Very small, simple prokaryotic microorganisms that are found almost everywhere. Some bacteria cause disease in humans and animals, while others aid natural bodily functions and are beneficial to health.

Bacteriophages (phages) Viruses that infect bacteria, hijacking their internal machinery to replicate. Most bacteriophages are lytic, causing the infected bacterium to explode and releasing progeny phage into the environment. Lytic bacteriophages are thus suitable for phage therapy by destroying the bacterial cell. Other bacteriophages are lysogenic and integrate their viral DNA into the host's genome rather than killing the bacterium outright.

Biobank (biorepository) A collection of biological specimens or electronic data records of specimens. Biobanks can also store associated health data. Material collected in a biobank is used for research purposes.

Biomass The total mass of all living organisms, or the set of organisms present in an ecosystem. A biomass includes a variety of species: producers, consumers, and decomposers. Their total mass is usually expressed as dry weight or, more accurately, as the carbon, nitrogen, or calorific content per unit area.

Biosphere The thin film of living organisms and their environments found at the surface of the Earth. The biosphere includes all of the environments capable of sustaining life above, on, and beneath the surface of the Earth and the oceans. The biosphere is composed of the entire hydrosphere and portions of the atmosphere and outer lithosphere.

Commensalism An interaction between two animal or plant species that live together. In a commensal association one species (the commensal) benefits from the interaction while the other species is not significantly affected with either harm or benefit.

Culture The cultivation of cells under controlled laboratory conditions. Cultures must provide a source of energy and raw material for biosynthesis. Culturing can identify an organism's ability to grow on certain media. Cultures can be used in the identification of species.

DNA (deoxyribonucleic acid) The genetic material of an organism, consisting of a double-stranded polymer of sugar 2-deoxyribose, phosphate, and purine and pyrimidine bases. The sequences of the bases found in DNA—guanine, adenine, cytosine, and thymine— carry and specify genetic information. DNA can undergo replication, for the purpose of propagating and preserving genetic information between generations. DNA also undergoes transcription, a process in which the genetic information from the DNA is expressed to synthesize RNA and, ultimately, the different components of a cell.

Enterotype Three distinctive categories for grouping the kinds of gut microbiome based on the comparative distribution of different kinds of bacteria when all of the available genes in a fecal matter sample are analyzed. The dominant genus defines the enterotype. In some people the genus *Bacteroides* is most abundant, in others *Prevotella*, and in others *Ruminococcus*.

Epigenetic These are heritable genetic changes that are not the result of changes in DNA sequence. Gene imprinting is an epigenetic phenomenon in which the expression of a gene varies depending on whether it is inherited from the mother or father.

Eukaryote An organism with a distinct nucleus that contains the genetic material of its cells.

Extremophiles These microorganisms can live in environments that are considered extreme for human life. Most extremophiles are prokaryotes belonging to the domains Bacteria and Archaea. They survive in ecological niches such as at high or low temperatures (160–230°F or 70–110°C; 32–50°F or 0–10°C), extremes of pH (acidic or alkaline conditions; pH below 4 and above 9), high salt concentrations (above 20% salt), and high pressure (above 300 bars or 290 atmospheres). Since the mid-1980s extremophiles have been objects of basic research and innovative biotechnology.

Fungi This is a kingdom or a large group of eukaryotic organisms that includes microorganisms such as yeasts and molds. The kingdom, Fungi, is separate from plants, animals, and bacteria. Fungi form a single group of related organisms, named the *Eumycota*, that share a common ancestor (a *monophyletic group*). One distinctive feature of fungal cells is that their cell walls contain chitin.

Genome The entirety of an organism's genetic information.

Genome-wide association study (GWAS) A type of genetic study in which a large number of genetic variants from different individuals are examined to see if any variant is associated with a particular trait. The number of specimens genotyped provides sufficient power to detect variants of modest effect.

Genotype The genetic makeup or particular allelic configuration of a cell or organism that expresses a particular phenotypic characteristic or trait.

Germ theory of disease In the late 1860s, the French chemist and microbiologist Louis Pasteur proposed the idea that microorganisms cause disease. This theory led to fundamental breakthroughs in modern medicine and clinical microbiology including significant innovations such as antibiotics and other medical technologies.

Koch's postulates A set of widely accepted criteria for determining a causal link between a particular bacterium and a particular disease. German physician Robert Koch used these criteria to identify the bacteria responsible for anthrax, cholera, and tuberculosis, among other diseases.

Lytic enzymes (lysins) The enzymes produced by bacteriophages (i.e., phages) that degrade the bacterial cell wall.

Metabolomics The comprehensive, qualitative, and quantitative study of all the small molecules in an organism.

Metagenomics The study of genetic material extracted directly from an environmental sample and the research techniques that allow investigators to characterize a microbial community or its members. Metagenomic studies aim at developing an understanding of transorganismal behaviors and the biosphere at the genomic level.

Microbiome All of the microorganisms (i.e., bacteria, viruses, fungi, and protozoa) and their physical interactions (pathogenic, commensal, and mutualistic) in a particular environment. The human microbiome refers to the population of microorganisms that live on and inside the human organism. The term *microbiome* is a combination of "micro," from microorganism, and "biome," a major community of plants and animals having similar life forms or morphological features and existing under similar environmental conditions. The term *human microbiome* refers to the population of microorganisms that live on and inside the human organism.

Microflora The microorganisms that occupy a particular ecosystem, either a plant or animal host or a single area of a host's anatomy.

Mutualism A relationship between a microorganism and its host in which both organisms are benefited by their interaction.

Nitrogen fixation The chemical or biological conversion of atmospheric nitrogen (N_2) into compounds that can be used by plants and become available to animals and humans. Microorganisms play a critical role in nitrogen fixation, as certain bacterial and algal species convert nitrogen gas (N_2) into ammonia (NH_3) that can be used by plants. Some nitrogen-fixing microbes establish symbiotic relationships with higher organisms, such as leguminous plants and termites.

Parasitism The symbiotic relationship in which one partner benefits and the other is harmed. This is the type of relationship that we most commonly imagine when we consider bacteria and viruses.

Pathogen A microorganism that causes disease in plants or animals.

Penicillin Penicillin is a substance released by the mold *Penicillium notatum*. In 1928 Alexander Fleming discovered that this mold kills bacteria. Fleming's recognition of the antibiotic power of penicillin did not lead immediately to a useful therapeutic product. Developing the technology for using penicillin as an antibiotic required many years of additional work at Oxford by Austrian biochemist Ernst Chain and Australian pathologist and pharmacologist Howard Florey.

Personalized medicine This is a model of medicine that tailors treatments to each individual patient. Personalized medicine aims to use information on a patient's individual genetic makeup and genetic information about the patient's disease to inform the physician of the most efficient form of treatment for that patient. For example, certain subgroups of cancer have distinctive genetic profiles. Also, certain genomes make some individuals more or less likely to respond to particular drugs. Both genetic information on the disease and genetic information on the patient can allow physicians to customize the selection of the most effective treatment protocol for a particular patient to maximize health benefits and minimize harmful side effects.

Personalized nutrition This is a model that tailors nutritional recommendations to each individual patient. Personalized nutrition aims to use information on a patient's individual genetic or physical makeup to customize a diet for a specific patient and allow the patient to experience maximum health benefits.

Phages See *Bacteriophages*.

Pharmacogenomics The study of interactions among drugs, the genome (the complete set of genes in an organism), and the proteome (the complete set of proteins encoded by an organism's genome).

Phenotype An organism's combined observable characteristics and traits. An organism's phenotype is a product of gene expression as well as environmental factors and the interactions between the two.

Prebiotic An indigestible food ingredient that beneficially affects the host by selectively stimulating the growth or activity of one or a limited number of bacterial species already established in the colon.

Probiotic Live microorganisms that, when consumed in adequate amounts, confer a health benefit on the host.

Prokaryote An organism that lacks a true nucleus or any other membrane-bound organelle.

Protozoa A diverse category of unicellular eukaryotic microorganisms.

Pyrosequencing A method of sequencing DNA that involves the detection of released pyrophosphates during DNA synthesis. In a series of reactions, visible light is generated that corresponds to the number and identity of incorporated nucleotides. This process allows the entire genetic sequence of an organism to be determined.

Ribosome A part of a cell that synthesizes proteins through messenger RNA and transfer RNA

RNA (ribonucleic acid) A nucleic acid present in all living cells. It acts as a messenger by carrying instructions from DNA for controlling the synthesis of proteins.

Sanger sequencing Also called dideoxy sequencing or chain termination, the Sanger method mimics the natural process of DNA replication. It incorporates the use of dideoxynucleotides, which, when integrated into a sequence, prevent the addition of further nucleotides. The process was developed in the early 1970s. It was one of the first methods for efficiently sequencing DNA.

Shotgun sequencing A method for determining the nucleotide sequence of a large stretch of DNA. It involves randomly sequencing a genomic library (q.v.) derived from it that contains clones with overlapping DNA fragments. It uses powerful computers to assemble the sequence data into a continuum.

Spontaneous generation A theory regarding the origin of life from inanimate (nonliving) organic matter. Pasteur discredited this theory during the 1880s.

Sterile The condition of a substance when it is free of all living microorganisms. During the late nineteenth century, English surgeon Joseph Lister, influenced by Pasteur's study of microorganisms, incorporated antiseptic methods into surgical operations by keeping wounds sterile and cleaning surgical instruments.

Symbiosis An interaction between individuals of different species (symbionts). The term *symbiosis* is usually limited to interactions in which both species benefit.

Thermophiles An organism, especially a microorganism, that can tolerate high temperatures. Thermophiles grow optimally at temperatures above 45°C.

Vaccine A suspension of bacteria, viruses, or proteins injected to produce an immune response and protect against infection by pathogenic microorganisms. Vaccines have been developed since the beginning of the nineteenth century. Our understanding of vaccines and their scientific mechanism derives largely from microbiome and microorganism research.

Virus These microorganisms have a very simple structure and often consist only of DNA or RNA with a protein coat. They lack some of the chemicals and enzymes needed for replication. To reproduce, they infect other living cells and take over the infected cell's internal "machinery." They are very tiny and not visible with an ordinary microscope but only through electron microscopy.

Volcanism The movement of magma and its gases from Earth's interior into the crust and to the surface. Volcanism is responsible, in part, for global temperature regulation and carbon dioxide levels.

Weathering Physical and chemical disintegration of earthy materials due to atmospheric agents. Weathering is partly responsible for global temperature regulation and carbon dioxide levels. The rate of weathering is highly influenced by microbes on the Earth's surface.

INDEX